*Dispatches from the
End of Ice*

# *Dispatches from the End of Ice*

ESSAYS

BETH PETERSON

TRINITY UNIVERSITY PRESS
San Antonio, Texas

Published by Trinity University Press
San Antonio, Texas 78212

Copyright © 2019 by Beth Peterson

All rights reserved. No part of this book may be reproduced in any form or by any electronic or mechanical means, including information storage and retrieval systems, without permission in writing from the publisher.

Book design by BookMatters
Cover design by ALSO
Maps and illustrations by Maya Blue
Author photo by Bobbie Peterson

Frontis: On July 12, 2011, the crew from the U.S. Coast Guard Cutter *Healy* retrieved a canister dropped by parachute from a C-130, which brought supplies for some midmission fixes. NASA/Kathryn Hansen.

ISBN 978-1-59534-899-9 hardcover
ISBN 978-1-59534-900-2 ebook

Early versions of these essays were published in *River Teeth* ("Glaciology"), *Passages North* ("Driving Wyoming"), *Newfound* ("Lost: An Inventory"), *The Pinch* ("Speed of Falling"), the *Ocean State Review* ("Baffin Island"), *Flyway* ("Theory of World Ice"), the *Mid-American Review* ("Wittgenstein's Cabin"), *Terrain.org* ("Finding Atlantis"), and *Post Road* ("Journey to the Center of the Earth").

Trinity University Press strives to produce its books using methods and materials in an environmentally sensitive manner. We favor working with manufacturers that practice sustainable management of all natural resources, produce paper using recycled stock, and manage forests with the best possible practices for people, biodiversity, and sustainability. The press is a member of the Green Press Initiative, a nonprofit program dedicated to supporting publishers in their efforts to reduce their impacts on endangered forests, climate change, and forest-dependent communities.

The paper used in this publication meets the minimum requirements of the American National Standard for Information Sciences—Permanence of Paper for Printed Library Materials, ANSI 39.48-1992.

CIP data on file at the Library of Congress

23  22  21  20  19  |  5  4  3  2  1

*for Kate Northrop,*
*Elizabeth Chang,*
*and S. A. Stepanek*

## CONTENTS

### BEFORE
Baffin Island    1

### DISPATCHES FROM THE END
Theory of World Ice    19
The Philosopher's Cabin    38
Driving Wyoming    59
Lost: An Inventory    79
Glaciology    97
About the Collection    113
Pauling's Core    134
The Iceberg Proposal    146
The Speed of Falling    160
Cairns    178
To the Center    199
On Time    220

### AFTER
Atlantis    235

Notes    251
Acknowledgments    267

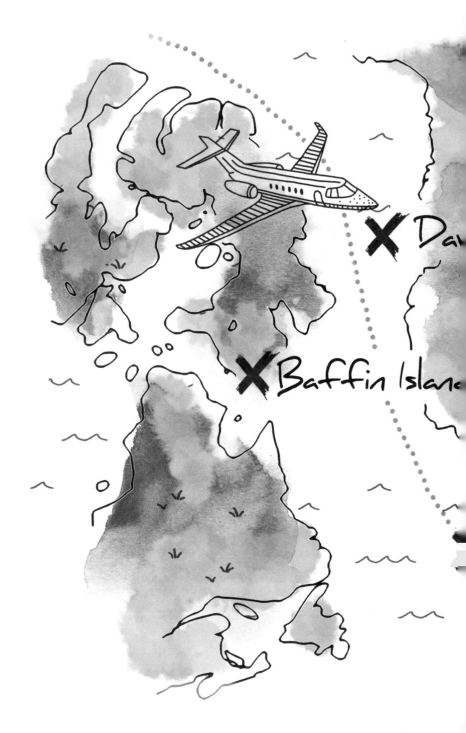

# BAFFIN ISLAND

38.9517° N, 92.3341° W

## THE AMERICAN MIDWEST

A few months after I leave the windswept plains of the Wyoming West for the last time, a flat, white document box comes for me in the mail. When I see the package resting against my front door, I will wonder for a moment if it's an old passport finally returned or maybe my university diploma. I pick up the box and, still standing on the porch—tall, narrow blades of bluegrass edging the boards, springing up through the wooden-slatted floor—carefully slide it open with a house key. It's not a passport, a diploma, or anything else I've been waiting for. Instead, out of that box slips a small, hand-tipped map, dated 1750.

The map is of Norway, but not the Norway I know. There are no country boundaries between Norway, Sweden, and Finland, only lakes, mountain ranges, and rivers. Regions are separated by color and by tiny dots, bordering the edges. Letters in the place-names—Berghen, Gothland, Stavanger—are fine and carefully printed, some closer together, some farther apart, the lengths and angle of *e*'s and *t*'s varying based on their size and position. The sides of the map are crosshatched in black ink, but the rest of the map is painted in faded yellows, blues, and greens. There's a water spot in one of the corners, then just above that, the name of the mapmaker, Robert de Vaugondy, and his own handwritten note in French "with privilege."

For almost a year before that box arrived in the mail—a gift, it turns out, from my friend and former professor, Kate—I'd been thinking about maps. I'd hung maps around the tall, white walls of my apartment; I'd watercolor painted a five-foot-wide outline map of the world in vivid blues and purples; I'd placed an old wooden globe that had once been my father's in the center of my desk, traced its continents onto scraps of paper. I'd even contacted an acquaintance from college who had become a cartographer and asked him if he could tell me how contemporary maps were made.

It had started with a question from a friend. I had intended to write a short collection of poetry in those days in Wyoming, but all my poems had been turning into scenes: nonfiction rescue scenes. At first, those scenes were traditional rescues—bystander saves boy from near-drowning in local lake; family of four and eleven cats escape house fire—but gradually they had become stranger and more unrecognizable: deer jumps through back window of 1998 gold Toyota Tercel; mobile home blows off semi-truck bed in Virginia Dale, Colorado; two men struck by lightning under the same flowering willow. After reading a draft of one of these scenes, a friend had penned a single note on the bottom of my page. "Where," he asked, "is the rescue?"

I read that note and then I read it again; I walked around, it seemed, for days, along the high mountain plain bordering the city where I lived, trying to pinpoint an answer to my friend's question. I knew that with his *where is the rescue?* my friend was asking whether I was actually writing rescue scenes if the people in the scenes don't get rescued. In the end, though, it was the other part of that double-meaning "where" which always eluded me. The thing I needed to know *was* where; where in physical space—in what mapped place—would we be safe?

60.6153° N, 64.5256° W

## THE NORTHWEST PASSAGE

Norse mythology tells a story about a vast chasm and about a rescue. The chasm was famed to be the entrance from one world to the next, the exact spot where the earth originated and where, in the end of all days, it would disappear. As the tale goes, the cosmos was first made of three parts. To the north was the homeland of ice; to the south was the homeland of fire; and between them was a great gap called the Ginnungagap. Described by one writer as the "chaos of perfect silence" and by another as the "yawning emptiness," the Ginnungagap represented nothingness: an immense primordial void.

Still, this void didn't stay empty forever. Over time, the land of ice and the land of fire began to move toward each other, gradually overtaking the Ginnungagap. Then, one fateful day, sparks from the land of fire and a frozen river from the land of ice finally met. The fire's heat warmed the ice until there was frost and then melting frost, and then the melting frost transformed into a frost giant and into a cow; the cow's milk was the giant's food; the cow's food was the salty ice.

It wasn't just ice though. Caught beneath all that ice was a man. That first day, as the cow licked the ice around her, the man's hair appeared. On the second day of licking her hot tongue against the frozen water, the man's head appeared. On the third day, the whole man was freed, cracked out of the ice like a nut from its casing. It was this man, Buri, whose great-grandsons would one day slay the giant, flinging forth from the giant's frosted body the stars, the sky, and the earth as we know it.

61.4905° N, 7.6006° E
## SKJOLDEN, NORWAY

I had not yet heard of the Norse myths when I first traveled to Norway, though perhaps I should have. Sometime before, when my Norwegian grandfather moved several states north—from a two-bedroom apartment he could no longer keep up in the South to a studio in the Midwest, just a few miles from my parents—he'd given me a stack of his old books: Dante's *Inferno*, *Journey to the Center of the Earth*, *Faustus*, and then one entitled *Norse Stories*.

"I think you'll like these," he told me, pointing toward a cardboard box on his old tan loveseat one afternoon, shortly after he moved in, before he'd even hung pictures on the walls. I cracked open the lid of the box; the ten or eleven books inside were all several decades old. There were a few worn paperbacks. Mostly, though, there were hardcover books, with slightly yellowed pages but still-crisp bindings. The first book I took out of the box was that book, *Norse Stories*; it was a blue-green hardback, with the title printed in circular letters, just above a pattern of vines-becoming-dragons that wrapped around a single sword.

I brought the books home but did not read them, not before I left for Norway. I was planning then to be in Scandinavia for just one summer: to explore a new place, to trek into different air, to live for a while outside of my own language and uncertain attempts at navigating a career and life and relationships. When I got to

Norway, I climbed mountains; I kayaked through snow-fed waters along the long green shoreline; I hiked the edges and then the interiors of Norway's wide fields of glacial ice—which, glinting in the high summer sun, even in the distance, seemed limitless—and I knew already in those first months that I would be back.

56.5110° N, 3.5156° E
## THE NORTH SEA

As early as the fifteenth century, cartographers set out to map the Ginnungagap. Early maps placed this supposed locus of creation somewhere between Greenland and the Atlantic coast of Canada. The gap, it was then believed, proceeded from a vast sea that circled the earth; the gap was a borderland between that outer ocean and the inner one, the ocean that wrapped around, lapped up onto the known world. Later maps pinned the Ginnungagap to the Davis Strait, home of the Northwest Passage, or to the southwestern tip of Baffin Island, with its fierce, fifty-foot tides, its ice-capped mountains, and its polar bears and caribou. In a 1606 map Icelander Gudbrand Thorlaksson located the Ginnungagap between Greenland and modern-day Ireland. In a 1507 map Johannes Ruysch—likely roommate of the artist Raphael—cast the Ginnungagap into the North Sea, as if to say *don't travel beyond this point.*

The first map I find of the Ginnungagap is a 1636 rendering by Thordur Thorlaksson, Gudbrand Thorlaksson's son. The map is preserved in a digitally archived book, with inked-in images, dark and delicate, of a bear, a fox, two men standing side by side on the top of that map, and mountains—drawn as puffs of smoke—circling a central landmass. In the place, though, where there's supposed to be the Ginnungagap—on the southern side of

Davis Strait, just below Baffin Island—a woman's hand obscures the image. It was an inadvertent copy, no doubt, perhaps quelling a gush of wind from an open window, bumping a green oval "save" button a moment too soon, or scanning one page-view while meaning to stop, turn the book, and scan another. In any case, three of the archivist's fingers are caught in that image, covered in pink silicone wraps; the other two and a large diamond ring are unwrapped but still bear across the page, each finger perfectly manicured, French-tipped nails pointing east.

41.3114° N, 105.5911° W
## LARAMIE, WYOMING

In one of the rescue scenes, I'm hiking with friends along a high mountain pass; it's summertime, but it's just begun to snow. We make steady but careful progress, one foot in front of the other, until we can't move forward anymore. Though we've been walking in twos until that point, suddenly all seven of us are there at the same bend. While we've been walking in all that falling snow, none of us has realized that there's a sheer cliff just beyond and the fog has come in thick behind; there's no way go to back and there's no clear path ahead.

In another rescue scene, I am trying to pick up a man from the bottom of a pool. It's early evening. The walls and the ceiling above the pool are painted white, but the dark floods in through the high poolside windows; it reflects on the water. It's the only reflection on the water; the pool deck is nearly empty; there's no one watching from the bleachers, no one standing outside the locker rooms, no one walking by in the long white-tiled hallway.

The man lying completely still at the bottom of the pool is my lifeguard teacher; he has instructed each of the eight students in my class to dive first for a penny, then for a red plastic ring and, finally, for his body; he has told us to hoist each of those things through the aqua-blue water and up, onto the rippled concrete. The other students have mostly already finished and gone; one other girl and I are the only ones who remain. The lifeguard teacher has blown his whistle and then descended, fifteen feet to the bottom. It's my turn to follow.

I do; I dive in a single clear motion, breaking through the water, going three feet then four feet then five feet down. Soon I'm seven then eight and then ten feet into the water, but as I see the body below me, it's not like the penny and it's not like the ring; it's pale and out of focus. I reach for it once and then twice but touch nothing; I come up for air and then try again. No matter how far down I dive, I cannot make my way to him.

In a third rescue scene, I'm back at the pool, except this time the pool is my white-walled bedroom and the water is the humid air and I can't tell if I'm sinking or treading water, only that it's the middle of the night and I'm twenty-five and nothing I know is certain. Outside, sleet is coming down hard, with no sign of stopping.

69.0579° N, 62.8628° W

## THE DAVIS STRAIT

The Ginnungagap isn't empty in every version of the Norse myths. One of those tales describes cold winds from that chasm transforming mist from the North into huge blocks of ice. The blocks of ice thundered as they fell into the gap and became rivers—there in the chasm—rivers of frost that flowed to meet the fire. In Snorri Sturluson's *Prose Edda*, the thirteenth-century source for the Norse legends, the fire is personified, a character at the very edge of the land brandishing a single flaming sword.

Another version of the story adds that the frost giant wasn't just "slain"; he was thrown off a high cliff by Buri's great-grandsons, back into the Ginnungagap. When the body of the giant landed, so much blood pooled that it drowned nearly everything around it. As the frost giant lay there lifeless, from his blood came the seas; from his skin came the land; from his hair came the trees; from his bones came the mountains; from his skull came the sky; and from his spilt brains—lush and gray—came all manner of clouds, straight out of the breach.

In an art exhibit entitled *Ginnungagap*, after the Norse myth, Swedish visual artist Sigrid Sandström created twenty-four multimedia paintings of varying sizes of the same icy landscape depicted in the *Prose Edda*. The paintings, some watercolor, some acrylic, were shaded in deep blues, blacks, and purples, sharp shards of ice and rock bordering a rift in one, softer rounded edges of snow spilling into a valley in another. Along with the paintings were two short films, playing on repeat. In one of the films, a man planted a black flag in an otherwise starkly white landscape. In the other film, a buoy propped up another small flag as it floated through the middle of a choppy blue sea. At the end of the installation, the man returned to take out the flag he had planted, and the buoy eventually drifted offscreen.

"When a man plants a flag," noted art critic Jen Graves, in an article about the exhibit, "he imagines that his ownership radiates outward from his hand like a nuclear blast, touching what has become his. But he does not make a home here. Sandström suggests that ownership is a poor substitute for knowledge—and that representing an object or a place in art is like staking a claim in inhospitable land."

79.2138° N, 14.6489° E
## SVALBARD, NORWAY

By the late 1600s mapmakers had begun to question their own renderings, "to struggle," as one historian put it, to reconcile the geography of the Old Norse world with contemporary findings. Eventually even the word "Ginnungagap" began to change in meaning—from the sacred creative space at the world's origin to any mighty or "chaotic chasm"—and Ginnungagaps came to mark on maps maelstroms or whirlpools and then the ocean itself. In the last maps that show the Ginnungagap, the word began to represent "holes" or "edges" where the world of people didn't perfectly fill the world of fire and ice.

By the 1700s the Ginnungagap and even the idea of the outer ocean had been framed in the topography of fiction. Explorers had not found the entrance to other worlds or other oceans on any of their voyages, across sea or across land, and the art of mapmaking had gained a different focus and sort of precision. With a clearer knowledge of longitude and a more scientific attention to tracing observational on-the-ground information, European mapmakers turned away from the imaginative and speculative elements present in prior maps and toward mapping, as one historian notes, "as utility."

In 1757, only seven years after making the map that came in that flat, white box to my porch, Robert de Vaugondy and his son published *The Atlas Universel*, a set of 108 maps that were noted not only for having more accurate latitude and longitude marks but also for citing in a preface their geographical sources. By this time, maps had begun to claim objectivity, operating, as theorist Anne McClintock puts it, as "a technology of knowledge that professes to capture the truth about a place in a pure scientific term, operating under the guise of scientific exactitude." Vaugondy and his son's maps of Scandinavia and other places were sent to more than a thousand subscribers initially but then circulated even beyond those hands.

By the time I search for "Ginnungagap" on the world map—250 years after Vaugondy—I find that the word now only denotes a single icy valley on the Norwegian Arctic island of Svalbard, dipping deeply between a mountain and a fjord.

60.3913° N, 5.3221° E
## BERGEN, NORWAY

When I first traveled to Norway, I carried with me everywhere a blue laminated notecard on which I'd penned the addresses, approximate locations, and best routes to anywhere that seemed important: the airport, the guesthouse where I was staying, the bank, the U.S. embassy, the store, the post office, even the national park where the Norwegian glaciers are. I had no phone with me in those days and no way to get online except one shared dial-up computer that cost four or five dollars a minute; it seemed prudent to always keep locations on hand.

I brought the blue card back my second summer in Norway and my third, but I realized even then—taking out the card my first day back that third year—that the place had changed. The bank and tourist information stops had both moved somewhere down the street; bus and ferry route numbers were different. The village where I stayed—once remote—now had cruise ships arriving five or six times a summer. When I returned to the glacier, there was so much less snow and ice that I thought at first that I had boarded the wrong bus.

My sixth summer in Norway, I searched for a map of Baffin Island. What I stumbled onto instead was a rash of recent news. Researchers led by Gifford Miller from the University of Colorado had been studying mosses on the island, carbon-dating the plant life that they were finding under the island's receding ice caps. Not long after beginning their study, the scientists began to uncover a strange phenomenon; as the ice melted, there was a fifty-year period when plants at different altitudes seemingly all froze at once, defying usual expectations for higher or lower freeze points dependent on altitude.

The time period, Miller and the rest of the researchers recognized, must have been the beginning of the Little Ice Age, at the end of the thirteenth century. After studying computer climate models and environmental disruptions at that historical moment, the scientists made a further realization; the abrupt onset of the freeze suggested something else important and yet undiscovered: the ice was caused by something swift, something catastrophic.

The Ice Age wasn't a fluke or a cycle, Miller hypothesized; tiny particles of sulfur suddenly blocked out a portion of the world's sunlight. In Miller's view, these tiny particles—the ones that caused many of Europe's glaciers—were slung into the atmosphere by fifty years of erupting volcanoes.

There was a photo accompanying several of the articles. It wasn't of volcanoes or even glaciers though. Instead, it was of Gifford Miller—ruddy face, short gray hair, black waterproof trousers, hiking boots, and faded red shirt, one button anchoring the collar—out on Baffin Island. Miller was crouched close to the ground; he held a clear plastic bag in one bare hand and combed a small rocky outcrop with the other. Behind the small rock island where Miller perched was a long expanse of glistening, wet snow, reaching all the way to a low bank of clouds and then the end of the frame.

34.6037° S, 58.3816° W

## BUENOS AIRES, ARGENTINA

Jorge Luis Borges famously tells the story of a single map that gets so large that it begins to cover the whole world. "In that Empire," Borges writes, "the Art of Cartography attained such Perfection that the map of a single Province occupied the entirety of a City, and the map of the Empire, the entirety of a Province. In time those Unconscionable Maps no longer satisfied, and the Cartographers Guild struck a Map of the Empire whose size was that of the Empire, and which coincided point for point with it."

The short story that map appeared in was a single paragraph long and was published in 1946 as part of a larger piece called "Museo" (in English, "Museum").

In 1982 Umberto Eco used the complete story as an epigraph for his own fanciful essay, "On the Impossibility of Drawing a Map of the Empire on a Scale of 1 to 1," which argues that the one-to-one map of the world must not be plaster cast, lest it pave the whole world, and if the map is suspended in the air, the subjects creating the map would be unable to move because every movement would alter the map. If the map is carefully folded, there could be a dangerous clump of all the folding people, adds Eco, and if it's opaque, it would block out the sun, thus changing the land that the map is trying to signify.

In 2006 Neil Gaiman likewise seemed to use the story as the basis for another very short story about a Chinese emperor, aptly titled "The Mapmaker." "One describes a tale," he begins, "best by telling the tale. You see? The way one describes a story to oneself or to the world, is by telling the story. It is a balancing act and it is a dream. The more accurate the map, the more it resembles the territory.... The tale is the map which is the territory," he continues a line later. "You must remember this."

Forty years after telling the story of the map-the-size-of-the-empire, Borges took on a new project. He decided to translate the first section of the *Prose Edda*, the section of that medieval Icelandic work which details for readers the Norse Ginnungagap myth.

"En el principio fue el tiempo en que no había nada: ni la arena, ni el mar, ni las frías olas; ni abajo la Tierra ni arriba el Cielo, sólo Abismo," Borges writes in this translation, *La alucinación*. "In the beginning was the time when there was nothing: neither the sand, nor the sea, nor the cold waves, neither below the earth nor above the sky, only abyss."

I try to work my way through the text in my limited Spanish late one night, Borges's translation and a language dictionary on my lap, side by side. I get through a few paragraphs at a time, penciling in the words I remember, looking up the ones I don't, trying to follow the logic of Borges's words, the way he translates the images.

As I read, I can't stop thinking about the map-the-size-of-the-empire, though. I can't stop wondering if this translation somehow speaks to its reverse, to what happens when a map gets so small the places begin to slip off it entirely.

65.4215° N, 70.9654° W

## BAFFIN ISLAND, CANADA

Only one of my rescue scenes contains a map. It's a map of *Pilgrim's Progress* that I painted when I was eight years old. The map had been stuck on the top shelf of a closet in my parents' house along with some stories I had written and some drawings of my brother's; my father had found it when they were cleaning things out twenty-some years later, had brought it down and set it out on a dresser, maybe to look at himself, maybe so I would look at it.

In that map, a long brown road curves from one of the book's locations to another—the Tomb, the Great Woods, the Hill of Difficulty—each spot carefully painted and then outlined in black, sometimes with tiny characters drawn beside it. Random spots of green appear all over the map—trees perhaps—and my first name only is penned in thick marker at the top. Near the end of the road, the Palace Beautiful, painted in yellow, with a flag on top, two turrets, eight windows, and one door, is just around the bend from the Castle of Giant Despair, which is one large gray block.

In that rescue scene, though, what my eye focuses on is the very corner of the map. There the Celestial City is covered in water.

More than a year before I left Wyoming—a few months before I began to write all those rescues—my plane to Norway happened to fly over the Davis Strait and Baffin Island, the onetime site of the Ginnungagap. We were in the clouds but then the plane had dropped, and though we were high above the cawing of birds, it was as if we were among them. Just as we were approaching the island, we could see the sea and the soft edge of land, white stone cliffs, sharp and treeless, and ice—snow and ice—stretching across them, taut, like the walls of a staked tent or a wedding canopy. After each mountain, a single dip and then another, rising higher, falling farther. The sky was a few shades lighter than the water on the shoreline had been, but heavy still with the weight of dusk behind the double-paned airplane window, with the short slope of approaching dawn.

You see, though I could not predict what was to come, I felt instinctively in those moments above Baffin Island that if I ever landed there—in the space between the worlds—even with a map, I might not find my way out. I didn't yet know, however, whether this tremendous ache of possibility was a comfort or a weight.

# DISPATCHES FROM THE END

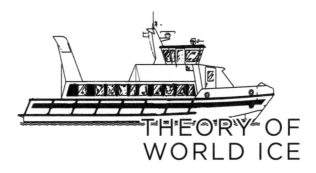

# THEORY OF WORLD ICE

On a cool day in late May, a friend and I take a boat heading north toward the largest glacier in continental Europe. My hair, though tied back, whips around me in the wind and so I cinch the hood of my rain jacket tighter. I feel the steady bob of the boat and can hear, even over the sound of the engine, the waves slamming into the hull and onto the rocky shore. The air is damp, more from the wind than the occasional drizzle, but there is still a short-range view: rounded rocks and small villages rising from the water—wood against stone—all framed in the low sky beyond the boat.

Outside on the deck where we sit, the wind hits hard and the water from the engines tunnels behind the ship in a wide white track, no matter what section of the fjord. The track slopes and angles, blurs into a peninsula and is redrawn. I follow with my eyes several hills of white water, each cresting for a few seconds before hurling then edging itself back into the sea, back into the geography of only memory.

When the boat took off from the Bergen docks, a dozen other people lingered on the deck wearing rain jackets and sweaters, some leaning on the boat's long railing, others milling about, looking at the water or the clouds or the city through the screens of their cameras or phones. A faded Norwegian flag flapped against its flagpole; the small orange rescue boat bounced up and down with the motion of the ship. Some people were smoking, others taking photographs of the city skyline: glass and concrete buildings, tall narrow houses, and the red tents that cover the local fish market getting smaller and smaller. Some people waved though no one on the shore waved back.

The boat continued on anyway, out of the rounded harbor, shouldered by rolling shoreline, by rows of shops and stone churches, by streets filled with houses and blocks of apartments, until it had left the city altogether. It moved past rocky strands circled by seagulls and small wooden cottages with tile or stone roofs, mostly crisply painted despite the water and the wind that must be constant. It blew its horn as a first and then a second smaller boat came into view but glided by them unaffected. It turned under an almost impossibly tall steel-cabled bridge, picked up speed, and cast out, into the open sea.

In 1894, outside under the stars, Austrian engineer Hanns Hörbiger had a vision. When he looked at the moon—its bright white surface shimmering in the night sky—he realized it was ice: the moon was made of ice. The galaxies were built by ice—great pieces of ice—that shattered into stars and planets as they fell. The deep matter of the universe, Hörbiger suddenly recognized, was ice; the cause for evolution was ice; ice was creating new worlds and destroying others.

A short time later Hörbiger dreamed that he was floating through space watching a pendulum swing farther and farther until it broke. That pendulum was gravity. "I knew that Newton had been wrong," Hörbiger would write, "and that the sun's gravitational pull ceases to exist at three times the distance of Neptune." From this vision and this dream came Hörbiger's *Glazial-Kosmogonie*, also called *Welteislehre*, or theory of world ice.

Central to Hörbiger's new theory was an alternate history of the universe where the solar system began when a massive star crashed into a much smaller waterlogged star. From the force of the crash, pieces of the smaller star were flung out into space. The water that star had been storing up eventually froze and became blocks of ice; the blocks of ice fell into a circle and formed that great spiraling galaxy which the rest of us know to be the Milky Way.

The boat we're riding on—a passenger ferry—is a narrow, very white catamaran. It's a two-story boat; there's an open area on the second floor with a few booths and tables and then, downstairs, luggage racks, a snack shop, and wide rows of plush seats, all facing forward, toward a computerized map of the voyage. The ferry is lined with windows and a few people follow the water through them, but most seem to talk or work or read or sleep.

The ferry is always part tourists, part locals. This day, there are maybe sixty or seventy passengers including a small group of schoolchildren, lots of older adults, and a few scattered young people like ourselves. We've taken the early boat—one of only two daily ferry departures to the small city of Sogndal—and it's a weekday, which usually means fewer travelers and more locals onboard; likely, my friend and I are the only ones making our way, this cold day, to the glacier.

The ferry passes under one high bridge, then a second, stops at one tiny port and another; as the degrees of landscape change from cityscape to rock island villages, the rest of the passengers leave the deck to go inside, one and then a few at a time. The trip is briefly in the open sea but mostly shielded by rocks and small masses of land that get increasingly larger. In Norway—unlike the landlocked places I've always lived before—the water acts as a sort of seasonal interstate highway, sometimes the only way in and out of certain villages and landscapes. When rocks or mud or early snows take out the single-lane roads that line the fjord in some of these places, boats become primary means of transportation, and often, even in the best of conditions, are quicker and more direct than traveling by bus or car on land.

I like riding the ferries; when I first got to Norway, my friends had a little red hatchback that we drove up and down the mountains, six of us packed into five seats or sometimes seven or eight of us if someone rode knees up in the trunk, against the lined glass of the sloping back window. The car was convenient, but it meant I rarely rode Norway's boats or busses; we moved either by foot or by car. The first time I discovered the ferries was a day that I was on my own. I was coming back into the country after time back home; that day, I made my way from one airport and bus to another, through the city, past the stone streets and wooden houses and green gardens and government buildings and market tents to the edge of the Bergen docks. I got on the ferry that day by myself and felt—as the waves swelled and we set off into them, a single boat on all the water of the world—awake for the first time in months.

All ice is made out of hydrogen and oxygen atoms in a frozen state. Still, there are several different types of ice—ice disks, ice pellets, shelf ice, candle ice, ice dams, aufeis or ice in streambeds—and various forms of all this ice. The most regular form of ice is ice$_h$ or hexagonal ice, when liquid freezes to below zero degrees Celsius. Ice changes, though, depending on temperature and pressure. The ice found on one side of a glacier in Norway may be different even than the ice found on the other.

Whatever the type of ice, when enough heat is absorbed, the individual particles gain energy, enough energy to separate, turning the ice into liquid water. I see a diagram of this online: the hydrogen as small red dots, growing like branches from circles of oxygen. The ice in the diagram is fixed, in neat, perfectly shaped hexagons. In the diagram for liquid water, however, the hydrogen branches are randomly placed, blue and red circles moving in dizzying motion.

Early into the ferry ride, I meet a South African graduate student named Luke. My friend strikes up a conversation with him as we board the metal gangway from the Bergen city center to the small ship. The three of us are the only ones who have carried expedition backpacks onto the boat and the only ones who will later choose to sit outside during the cold, several-hour ride. We can see the long chop of the white wake from where we sit on the floor, on the white painted second-story deck, our legs stretched out toward the water, our backs firm against the outside wall of the ship's cabin.

After some small talk about the ferry ride, Luke explained he was studying oceanography at the University of Cape Town, where he'd attended as an undergraduate and then returned. He'd had a gap year too, spent working as a crewmember aboard a small sailboat.

Luke was supposed to be traveling to Svalbard on this trip, an

Arctic island between Norway and Iceland that is mostly used for research into far northern plants and animals and climate. The scientist he was working with canceled at the last minute, though, so instead of taking one of the twice-a-week flights north, he had come to Bergen and now was floating around the country, seeing glaciers and watching World Cup soccer games in local pubs for six weeks until he went back to his life and his research in South Africa.

"Why did he cancel?" I ask Luke after chatting for a few minutes about his rearranged plans and scramble to find places to stay and things to see.

"I don't know," Luke says, shaking his head. "It happens sometimes."

He pulls his hooded sweatshirt over his short brown hair and broad thin frame as we talk, layering a fleece and a rain jacket the same way I have. Luke is friendly and, like several of my friends abroad, also charmingly boyish. He asks my friend and me about Chicago and Norway and why we're traveling here.

My friend rummages through her backpack and pulls out a small red-and-blue glass gnome. "I'm going to take pictures with this guy," she replies. Luke laughs.

For a while we compare notes about our travels: broken-down trains, missed flights, the expensive and bland food we'd all been eating, the best spots to camp and hike. When my friend eventually goes off to take photographs from some other position on the boat, Luke and I stay and sit quietly, eating sandwiches and watching the long expanse of sea distend and surge.

Norway's fjords—like its land—are glacial built, remnants of the last major Ice Age. The whole of Scandinavia was cast in those years in a sheet of ice, up to 3,000 meters thick over 6,600,000 kilometers of land. That ice sheet—the Scandinavian Ice Sheet—is said to have originated in Norway but then to have stretched throughout much of northern Europe, from Russia to the UK and Germany.

As the ice moved, it carved vast valleys and steep cliffs. When it retreated, many years later, sea filled in the valleys, creating clusters of long, deep fjords beside both Norway's low rocky shores and its high, barely inland mountains, steeped in spruce and pine. The land, scientists say, is still rebounding from the weight of the ice, rising from the water several millimeters a year.

Many years later another ice age came: the Little Ice Age, a period between the 1300s and 1800s when more snow fell in the winters than melted in the summers. Much of Northern Europe cooled during that period. The canals in the Netherlands are famed to have frozen then, as did London's River Thames, more than twenty times. In Dutch artist Abraham Hondius's 1677 oil painting of the Thames, people are shown walking and skating across the frozen river under a clouded sky. "Frost fairs" were held on the ice; one time an elephant was led across the river. In France, when the ice threatened to take over the Arve Valley, exorcists were brought in to call off the spirit of the advancing glacier. A forty-meter-high wall of snow came anyway.

"Professors were molested in the streets," Pauwels and Bergier write about the tactics of Hanns Hörbiger's world ice proponents. "The directors of scientific institutes were bombarded with leaflets, 'When we have won, you and your like will be begging in the gutter.' Businessmen and heads of firms before engaging an employee made him or her sign a declaration saying, 'I swear I believe in the theory of eternal ice.'" Other ice disciples were said to have shown up at traditional astronomy lectures shouting, "Out with astronomical orthodoxy! Give us Hörbiger!" Hörbiger's was a world that followed its own enticing logic. The movement created its own archives, its own genealogies, and even a newspaper, the *Key to World Events*; it was a metanarrative, Hörbiger believed, that could restart science on nonsectarian terms. And it was with no small sense of consolation that scientists, artists, and philosophers took up cosmic ice theory, not many years after the scientific revolution and not many years before much of the world would break out in a series of world wars over land and power and philosophical regimes.

"Modern Science seems to foster a desire for a final synthesis, a well-formed formula of the world that could eliminate the fragmentation of contemporary knowledge and its isolation within various academic disciplines," German scientist and natural historian Max Benzen wrote in 1934. "This metaphysical desire is expressed in two scientific ventures of the time: Albert Einstein's Theory of Relativity and Hanns Hörbiger's Theory of Cosmic Ice."

Both in Norway and around the world, for some time, the ice seemed limitless. In the 1880s—the same years my great-great-grandparents lived on the other side of the glacier—ice was cut in chunks from the edge of the Jostedalsbreen to ship all over Scandinavia for cooling.

An American, Frederic Tudor of the Tudor Ice Company, was the first to imagine that ice might be shipped and sold. On a visit to the Caribbean, Tudor was surprised by the stifling heat and wished, as he did back home in Massachusetts, he might cut a piece of ice from his family pond to cool himself or even just a drink.

In 1806, at the age of twenty-three, Tudor hired a boat, the *Favorite*, loaded it with ice, and set off for the island of Martinique, two thousand miles from Massachusetts. Eventually Tudor sent ice to Charleston, Calcutta, Savannah, Havana, and a host of other southern cities. Workers cut this ice from Fresh Pond, Doleful Pond, Spy Pond, Sandy Pond, Horn Pond, Spot Pond, and, during the years when a young naturalist named Henry Thoreau was living in the woods, from a sixty-one-acre lake in Concord, Massachusetts, called Walden Pond.

A couple of hours into the trip, the wind picks up again. I hold my hat on with my hand. The views from the boat become even more impressive, though, as the ride goes along, and I don't want to see it all through glass: the granite ridges rounding, then dropping straight off into the water's edge, the long patches of snow, waterfalls and glacial-fed rivers, small villages, and then even glimpses of the ice.

The Norwegian ferries seem to mostly run weather notwithstanding. I'd ridden on one, only two weeks before, when the open sea was so volatile that bags and drinks were flying off tables and people throwing up in wastebaskets.

"Please move to the center of the boat," the captain had finally announced on the loudspeaker in Norwegian and then English. At the next stop—a tiny village with little more than a few houses—half the passengers got off the ferry. I had stayed on for the whole ride.

This day, my friend goes inside, but despite the now steadily falling rain, Luke and I keep sitting outside. I pull my legs toward my chest and drape the plastic rain cover from my backpack over them. Just past the mountains, the clouds are low and thick, obscuring what's behind them, darkening even the early afternoon sky, making the familiar route look distant and strange.

The rain hits the boat sideways, pooling along the lower railing and running down the deck in dozens of streams toward the back of the boat. I readjust the hood of my rain jacket and rub my hands together under the rain cover; one of the streams hits the edge of my right shoe, then diverts around it.

"You know," Luke says after a while, nodding toward the fjord, "all this used to be ice."

I ask him to tell me more and he tells me about glaciers and about how carbon dioxide—from humans and other sources—is absorbed by the world's water, how he wants to know, but doesn't yet, what effect this has on plants and animals and the water itself. He's begun studying this in the Southern Ocean around South Africa and now wants to trace it north to the Arctic too. He explains that it all has an effect; even noise changes water: carbon dioxide, heat, ships, melting ice.

"The ice is compacted," he says, "something like clear plastic. If you were down at the bottom of a glacier, you could look through it like a window."

I nod along and then lean back against the boat's side and trace the wet white deck with my gloved hand. I watch the misted water, just beyond my feet, in motion, rolling. As I look out at the water, for the first time ever, I think of all the hydrogen bonds in that sea breaking—from ice to liquid water—and I wonder if they made noise as they did; I imagine the glacial ice caps that were once there, polished: smooth as pool balls, as pressed flowers.

In the early 1990s Norway's glaciers appeared to be growing. Though most of the world's ice was shrinking, in those years some of Norway's glaciers experienced higher snowfall than normal and actually began to expand. "Glaciers in Norway have begun to creep down from their mountain strongholds," a 1990 news article noted of Norway's Briksdalsbreen Glacier, "in apparent defiance of global warming."

I see a chart of glacier cumulative front variation, or changing lengths, prepared by scientists at the University of Bergen. In the 1990s—and, in some cases, as early as the 1950s—some of the glacier lengths did began to increase.

By the 2000s, though, every single one of Norway's glacier lengths had dropped, like most of the glaciers worldwide. At last count, France's Mer de Glace Glacier has retreated 2,300 meters. Norway's Rembesdalsskaka has retreated 2,000 meters. In a few months alone, the Briksdalsbreen glacier in Norway retreated 130 meters, far enough that it began to break off of the rest of the ice.

I first hear about Hanns Hörbiger and his theory of world ice in a lecture sponsored by my university. The lecture is put on by another department, but a friend sends me the dark, art deco–styled flyer with a note, "Ice: maybe you'd be interested?" On the left side of the flyer is a graphic of swirling cosmos in a black background; on the right the title "Counter-Science: The World Ice Movement's

Cosmic Visions and Its Rise to Fame (1894–1945)" and an image, perhaps a book cover, this time with an astronomical dial, yellow planets aligned but a triangular mass of particles spreading out from the sun.

My friend knows me well. Ironically, the scheduled lecture is canceled for the rare heavy dumping of spring snow in Missouri, but when the talk is rescheduled, I tell my students about it and let them out of class early so they can attend if they want. One of my students does attend; he, five other people, and I show up in a room built for forty or fifty, long rows of seats sloping toward a podium and screen in the front, toward a German scholar dressed in a turtleneck and jacket.

I take furious notes; I tape the talk on my phone. "How have I not heard of Hörbiger?" I ask my friend who sent the flyer when we go for a run on our favorite trail the next day, a layer of already-packed snow turning underneath our feet, the dusky light paling the leaf-bare trees we pass. "Well, I'd never heard of Hörbiger either," she replies, "though I guess I'm not as interested in ice."

I try to find an English translation of the *Theory of Cosmic Ice* and spend an afternoon in the stacks of our university library, looking for critical articles or historical newspapers that mention his name or cosmic ice. After hours of searching, though, I only find two new facts: first, that Hörbiger thought even the Northern Lights could be explained by ice—maybe cosmic ice dust, maybe, as one writer described Hörbiger's thinking, "distant glaciers reflecting the sun."

Everything else is in German, comes from sources that I'm not sure I can trust, or goes over material that the scholar had already mentioned in the Cosmic Ice talk. Everything, that is, except for one additional fact, a fact that I will take with me and not after that day forget: Hanns Hörbiger is said to have called his new world theory "the astronomy of the invisible."

We've made it through most of the Sognefjord—seen several of its icy plumes—when we round a bend and the ferry begins slowly aiming toward a landside dock, covered in a row of black, half-blown tires. It's Luke's stop; he's getting off the boat before we do, taking a railroad named Flam—one of the steepest trains in the world—to a high mountain village, then coming back the same way again. We walk Luke down to pick up his luggage and wave at him as he disembarks.

Soon we disembark too, hoisting our backpacks off the gangway and onto a still-wet dock as the *Fjord1* speeds away in the distance.

When we arrive at the glacier, it's too late to hike and so I camp overnight, in view of the ice but not quite at it, on the edge of a small campground. There are clean bathrooms, a kitchen, a washing machine, showers, and even wireless internet in the campground, but I pitch the tent as far away as I can from all this and even the other tents and caravans. I sit along a white-stone shore on a glacial-fed river and wash my feet and my face in the icy water. I lock the zippers to the tent with a luggage lock, realizing the ridiculousness of this and unsure what I'm trying to keep in or out but still doing it. I wake up to frost-covered ground and a tourist bus carrying a group from New Jersey.

The wind had settled into the valley again that morning; I could feel it from the blacktop parking lot where a local bus dropped me and two other passengers—teachers from California—off. It rustles the sleeves of my jacket, pulsating against my already chapped face. My friend had traveled farther on the evening before and so I was on my own this day, making my way toward the glacial ice; I could see it from the parking lot: wide ridges of snow in the distance, staggered along the crease between mountains and angling down, though not all the way to the base of the mountain, to the lake or the low forest between me and it.

"Mind if we join you?" one of the teachers asks as we get off the bus, pointing toward the steeply pitched trail ahead.

"Not at all," I reply.

There is a narrow break in the trees, just past the parking lot, and so we walk down a few dirt-carved steps toward it. The wind immediately cuts back; there are scraggly trees on both sides of the break and an ankle-deep partially frozen stream crossing under a few logs and pieces of squared wood. Just beyond the small clearing is the trail. As we begin down the stony footpath, I talk with the teachers for a little while about our lives and jobs and the different mountains that we've hiked. Eventually, though, they fall a few paces behind and I continue on in silence, pulled forward through the low canopy of dark branches by the steady sounds of the moving air and my own shoes striking ground. I walk over rocks and brush, past trees whose shadows are almost smaller than my own but glance still off the wet ground in a thicket of movement. I watch the shapes as I walk: thin rods of darkness interlocking and then suddenly splitting, letting in pieces of air and light. The path changes with them too, sometimes small pieces of gravel, sometimes bare ground, sometimes long stretches of slick rock.

I walk over several small streams, part ice, part water, hike up and down and around wind-washed boulders, some smooth, some patterned with winding orange and gray striations. I climb a single set of wooden steps onto a wide boardwalk that, the next time I visit the glacier, will be completely washed out, smashed up by spring flooding or a winter storm or maybe just the regular beat of wind through the valley.

I follow the pierlike bleached wood onward, up and up toward a long ledge of rock, toward spots of sun falling from the gossamer sky onto the snow-covered mountain in the distance. There is a metal sign about walking on the glacier at your own risk somewhere past the wood, and a field of rocks, mostly smooth, some small enough to hold in a single hand but others as wide as ponds or rivers. High on both sides of the expanse of rock are snowy hills—perhaps mountains, I'm not sure—treeless but mossy, ascending like a frozen sea swell, high enough I cannot see beyond them.

I pass quickly across level surfaces, steady my hands and feet along the smaller piles of rock. I listen to the path shift under the weight of my body. There are no animals and no other people, nothing breathing but the land ahead and behind and the teachers from California. The distance between me and them grows and I let it, though I stay always in earshot. As they fall farther and farther behind, I clamber over one small ledge and another until, finally, I'm there, at the farthest edge of the glacier.

Snow walls rise a few stories above the broad flat rock where I stand; they gather in apartment-sized drifts, blown like the foam cresting on waves, like sand on dunes. Cirrus clouds perch overhead, and a scattering of light gleams off the steep surface of the ice, fills it up like a glass. In places, that light is diffused—shades of sun swept up into thick currents of snow—but in other spots, the light is sharp and clear. Still, even from a distance and even in all that light, I can see that the ice is less like a mirror and more like a cratered moon, like the rounded, scarred underside of my own hand.

I move closer—five or ten meters—until I'm close enough to make out individual grains of ice, dense and compressed but still glinting. There's a tunnel ahead, wet and rounded, and sheer walls of snow, streaked with small bits of sediment, thousands of them. It's indigo—the blue inside that tunnel—not aqua, not robin's egg, not cobalt, not violet. It's urgently blue, luminescent. It's the blue of the fjord from a plane's window, the blue of an almost-night sky, the blue of the earth as viewed from some distant constellation.

It's a blue I've seen before, at night, in a Puerto Rican bay, when hundreds of dinoflagellates circled my kayak, flashing neon against a dusky sky. The sky at the glacier is not dusky; it's a hazy white, but the ice is the same: unexpectedly blue. And in the same way I thought of cupping my hand through the water in the ocean, here too I suddenly have an urge to touch the blue, as if I might take it in, might hold onto the color and the moment, absorb it into my own skin.

I walk forward and reach out my hand but at the last second stop myself and pull my hand back, remembering how even one warm body can change a landscape.

I step away from the glacier and walk a few paces down the mountain. Wind pulls at my hood and seeps through my thin jacket again. I can hear the pair from California now; they've moved closer and are taking photographs of each other and the ice. They're talking yet the noise of their voices is muted, even in the short distance between us.

It takes me a while to decipher what the other noise I hear is. It's steady and in the background, pushing against the now-loud wind, the sound of my new friends and the flapping of my own backpack and jacket against my shirt and my bare arms. It's hollow and constant and coming from somewhere deep below where I stand, the sound of something thrown against ice walls, pummeling through the ground, dashing against the granite.

There on the edge of the glacial ice is the sound of water rushing.

A few months before my first trip to Norway, I went on a boat to see other glaciers, that time to Alaska. It was a cruise ship then, and I was on it with my parents, my brother, his wife, and his in-laws. I was sleeping for ten days in a tiny, shared room, on a sofa that did not pull out. I was so cold that on the first stop of the trip, I had to buy another jacket. There were hot tubs and swimming pools on the ship—probably remnants of some other time when the boat sailed in a different season or a warmer place—but I never once saw them used. The towns the cruise ship docked at weren't even real towns but ports staffed by cruise ship employees eager to take money for excursions or mailed-in trinkets. I learned this one day walking past the port, a few miles to the real town, a town where there were no trinket shops and where a grizzly bear would later that day be reported roaming the streets.

What I did see from that ship were shards of ice, floating in the water like winter salt on a dark road. And one night, standing on the top deck, I watched a huge sheet of ice from the side of a glacier crack and hit the surface of the water, the snow around it rising like gunpowder. Later I'd watch a video of the ice breaking, or what I assumed was the same glacier breaking by the date and time stamp. In the background of that video, when all that ice came down, people clapped and cheered.

Forty years after its introduction, Einstein's theory would be almost universally accepted; Hanns Hörbiger's *Welteislehre* theory would be discredited by even its staunchest supporters. Hörbiger's ideas, in that time, had gotten tied up with politics and bad leaders and problematic philosophies, but most of all, despite Hörbiger's best efforts, the larger public didn't believe ice could ever be as important as Hörbiger surmised.

My friend will meet Luke once more, later in the summer, at a bar in Bergen for a drink and to watch a soccer match. I won't see him again, though I will see his name again and again in the many months to come as I look through articles and research statements on ice and $CO_2$, on blue whales in the Arctic and on global warming and carbon and oceans. I will think of him when I read, back home, six years later, a new oceanography study, which found that the places where glaciers melt into fjords may be the noisiest spots in the ocean.

I will replay too the video from the cruise ship of that glacier cracking, and I will wish later that I had told him about it. I will imagine, as I watch, the underwater shifting of ice, huge panes of ice, like windows, splitting off into hundreds of small pieces. Though I will not know what this would sound like in liquid water, I will reason that there still might be a brief moment of calm and then a terrible crash. And then—as I will remember a line from an essay I once read saying—*the world is made and unmade*. I will wonder if all that ice, all that noise, was simply absorbed or if for a few brief moments when the pieces of it fragmented and fell, if it became thousands of tiny stars, cosmic dust.

No ice breaks off the day I hike the glacier in Norway, though a few years later, in approximately the same spot where I that day stood, a massive piece will cave—or calve, as they call it—swallowing the parents of a young boy.

# THE PHILOSOPHER'S CABIN

On my third time biking past it, I finally saw the sign: WITTGEN-STEIN CABIN, 45 MIN. That sign—the sign I'd been looking for for nearly an hour—was situated several meters off the road, partway down a dirt tractor trail, and between an overgrown field and a camping park full of black-and-red-shuttered cabins and signs for other things: *ice cream, huts for rent, dining*. There was a stick figure of a hiker drawn on the sign and some laminated pieces of paper stapled onto the wooden post below it, but the sign itself was penned in faded black marker, in cursive—one letter running into the next, one word running into the next—and tacked beneath another larger sign for a different hike.

I tipped my bicycle into the high grass beside the trail and walked closer. One of the laminated papers was a Norwegian map leading toward, I assumed, the cabin, except that, besides the starting point—where I was currently standing—the map was merely a series of three black lines snaking through a yellow-and-green background with no distinctive landscape markers, map posts, or even labels. There was a laminated paper in English too, but on the English sheet there was no map, only a few vague directions—*Go straight then turn left. Follow the o-marked signs*—and a faded photo of a large, dead evergreen; a coral lake with mountains just behind it; and a metal pole, rising straight from the rocks over the water.

I walked a little ways back from the sign, looked up, and scanned the tree line. There were evergreens and pencil-gray mountains all around me, bands of snow lining their highest ridges, then just like that, fading into dark, dense swaths of forest, bisected only by the occasional trail or sheep path. There were ledges of rock here and there and a single bird that approached ground then lifted off, but there was no dead tree or metal pole and no cabin, not that I could make out. Even after several minutes of scanning, all I saw were mountains and more mountains, trees and more trees; I couldn't place the photograph in relation to the landscape.

I was living four or five miles down the road from the Austrian philosopher Ludwig Josef Johann Wittgenstein's cabin that summer, in a white three-story guesthouse in the very center of a tiny Norwegian farming village, Skjolden. Though cruise ships would begin docking in Skjolden's blown blue harbor one year later and stacks of shoreline condos and vacation homes would follow the ships, that summer—the summer I was twenty-five—Skjolden was still a relative unknown, marked by the sort of comfortable sameness that comes from a long and mostly quiet history.

My friends Darrell and Annette had bought the ten-bedroom guesthouse in Skjolden two years before. They had seen the not-quite-falling-apart house when they were camping at a high farm in the mountains nearby and had decided, perhaps against all reason, to buy it. It was in a perfect location: a hundred meters from the village's only hotel and shop, bordering a river, and steps from both the mountains and the fjord, with views of the berry farms and vineyards terracing the mountainsides. Just behind those green mountains that flanked the village were higher snowy mountains, still-vast glaciers, and four of Norway's national parks.

It was Darrell who had directed me which road to take toward Wittgenstein's cabin, and it was his old bicycle that I had ridden to get to the trailhead. "It's easy to lose your way up there," Darrell had warned me as we were standing around the kitchen eating breakfast that morning, watching the glacial-fed river just outside the guesthouse windows break against the rocky bank.

"I could run you up there in the car," he offered. "I've been there before, though it was a long time ago."

It was my second summer in Norway. The first—one year before—I had been mid-cross-country move from Chicago to Laramie, Wyoming, for graduate school. Darrell and Annette had just bought the guesthouse, and my own apartment lease was up. When he heard of my plans, my grandfather offered to pay for the plane tickets and sent me to the airport with a black-and-white photograph of his father and the name and number of a Norwegian relative, carefully keyed with the rounded letters of his old typewriter.

I did not call the woman whose name my grandfather had typed for me. I didn't have a phone that year—or for many years—in Norway, and the only time I found a pay phone was when I was waiting for a bus in an otherwise empty bus station several hours from where we lived. When I tried to call my parents, my three-hour calling card ran out after just a few minutes.

What I did that first year in Norway was fix up the guesthouse with Darrell and Annette—that and go on ice climbs, hike alpine mountains and glaciers, and pick wild strawberries right on the edge of the highway. I camped in high forested hills in the rain, and I wrote words on rocks then threw them into the sea. I lived in a small purple bedroom in the guesthouse with another American friend, Annie, a room where I was able to sleep almost instantly, despite the constant light falling through the sheer curtains.

One year later—after the trip and the move and returning to my grandfather's small studio with photographs of my own— my graduate program director, Beth, called to tell me that she'd rallied some summer travel funds for the department, that they were mine for the asking. I was back in my hometown already for the summer, looking for work. I walked into a blue-striped guest bedroom in my parents' house while we talked, pacing excitedly between the wooden dresser, the windowsill, and the overstuffed chair in the corner.

A few weeks later I flew from Minneapolis to Reykjavík to Oslo to Bergen and took a boat all the way back up the familiar coast. I stood on the top deck of that boat—a passenger ferry, really—as the white, red, and yellow steeply pointed houses and hilly stone streets and eventually even the Bergen docks went out of focus, replaced by narrow slopes of rock, then wider berths of green mountains and almost-open ink-blue sea. When I got to the guesthouse late that night, the candles were still lit, swaying in glass jars hung from open windows.

"What we cannot speak about, we must pass over in silence." This was the seventh and final proposition on logic in Wittgenstein's 1921 *Tractatus Logico-Philosophicus*, his most famous book.

Language, suggests Wittgenstein in the *Tractatus*, is ordered by logic; it has an underlying logical structure, and this structure determines the furthest limits—the margins—of what can be thought and therefore what can be said. Individual words in a language name objects, combinations of named objects make sentences, and sentences form paragraphs and ideas and treatises. Religion, aesthetics, ethics, the spiritual, the mystical, the metaphysical—all the things we can't pen in, all the things beyond the bounds of logic—are also, Wittgenstein argued, beyond the bounds of language. When we think about them, the words get caught in our throats; they disappear because they were never there; all that was ever there was, as Wittgenstein put it, "nonsensical." This inability to name, Wittgenstein wrote to his teacher Bertrand Russell in 1918, was not only the subject of his book but also "the cardinal problem of philosophy."

I didn't know this was the cardinal problem of philosophy when I went to look for Wittgenstein's cabin. I didn't know that the *Tractatus Logico-Philosophicus* was one of the most important philosophical works of the twentieth century, or that Wittgenstein had written it largely before his twenty-sixth birthday, in a remote two-story, wood-sided cabin on a high rock ledge overlooking a lake, the same cabin I was trying to find.

After leaving that first wooden sign, I hid the bike in the grass not far away and then spent the next several minutes looking for any left-hand turns. The directions had said *Go straight then turn left*, so I decided I'd go straight until I found the first possible left-hand turn and then I would take it. And if that wasn't the correct turn, I'd go back and take the next left until I found the one.

As I walked along the dusty path, I passed pines—live and budding—and birch, with their pocked trunks and rolls of papery bark; I passed a tall field of yet-to-be-harvested hay and a swamp where the water from the nearby lake had overtaken the land, trees and parts of bushes rising out of the water, reflecting in it. Though it had been overcast in the morning, the light had broken through the cloud cover now, had fallen onto the lake and settled into the gaps between the trees.

I walked over a narrow wooden bridge, past some low underbrush. Only a few minutes later I was almost to the edge of the forest, the first place the road forked left. The path wasn't very well defined, but it seemed to head in the direction the map had pointed and so I took it anyway. I walked ten or fifteen feet along that fork, on the edge of where the forest began to sharply slope upward, into the higher hills and mountains. After only two or three minutes of walking, though, the path stopped suddenly, walled off by a wide watery field with waist-high brown grasses. There were endless rows of birch and spruce, but there were no other left-hand turns or red *o*-marked signs in sight. The path had disappeared completely, as if whoever had laid the trail decided to go that far, but not beyond.

Unsure of the best move, I decided to leave the trail behind and to ford any seeming breaks in the grass on foot, hopeful that the trail was present, just grown over. I tried a first inches-wide break in the grass and then another and then stopped and looked up to get my bearings before repeating the process. For nearly half an hour, I alternated between circling various deer and sheep paths and gazing up at the mountain, trying to read its green slope like a grid.

Wittgenstein began work on his cabin in 1914, the same year that his teacher, Bertrand Russell, predicted Wittgenstein wouldn't outlive. Two of Wittgenstein's brothers had died, one by cyanide and one by drowning. Soon after, there would be the death of a third, by shotgun. Wittgenstein himself had a "nervous temperament"; wherever he was, he always wanted to leave, despite or maybe because of his brilliance. He had moved constantly, from England to Austria to Norway, looking for a career and a home and someone to publish his book. All these things were *a recipe*, Russell once told a friend about Wittgenstein, *for an early death*.

"I said it would be dark," Russell wrote of a conversation with Wittgenstein about another move, a move to Norway, "and he said he hated daylight. I said it would be lonely, and he said he prostituted his mind talking to intelligent people. I said he was mad and he said God preserve him from sanity. (God certainly will.)"

Instead of an early death, after traveling to Norway with a friend, Wittgenstein bought some land on the edge of a large glacial-fed lake near Skjolden and began constructing his cabin. It was a beautiful cabin and remote, built into the upper edge of the cliff, the only house on its side of the mountain and set a lake away from the rest of the village, with open views of the valley and the fjord. In the summer Wittgenstein would get to his land by rowboat, lodging in the village while the cabin was being built. In the winters after the cabin was finished, he snowshoed across the lake every week or few weeks for supplies. "I can't imagine that I could have worked anywhere as I do here," Wittgenstein wrote in 1936. "It's the quiet and perhaps, the wonderful scenery; I mean its quiet seriousness."

Some people in town called the cabin and Wittgenstein himself "the Austrian." Unlike the simple one-story, boxlike cabins popular throughout much of Norway, Wittgenstein designed his home to look more like a typical Austrian house with a second-floor balcony, a steeply pitched roof, and a wooden frame set atop a partially submerged stone and brick base. He constructed a system of metal poles and pulleys to haul groceries and other goods from the shoreline up to the cabin, many meters above, and then settled in to write his philosophical treatise.

The hike to Wittgenstein's cabin was the idea of my friends, Stephen and David. A few months earlier, David and I had run into Stephen while waiting for a train, the same train that routed just past my bedroom window on its way from one side of the city to the other, a train which sometimes rattled my pictures straight off the walls. It was a random Saturday night in May, cool and wet and quiet, like many Saturday nights in spring in Chicago, the only noticeable sound water sloshing under tires and draining down the streets into large metal sewers. David and I were going to the opera; as we walked up to the station, there was Stephen, standing out in the rain, leaning against the back of the building: tall, thin, wet blond hair. He was going to a play, catching the same train.

I had been the graduate teaching assistant for a course Stephen was taking a year prior. He was a philosophy major in an undergraduate poetry workshop; he had called me at home, at midnight, a few weeks into class, to ask me a question about an upcoming exam.

I picked up the phone, bracing for an emergency. "Hello?"

"Hi," he stumbled, "it's Stephen, you know, from poetry."

I told him the answer to the question he was asking but also that if he called the regular professor of the course that late, she might think about failing him. He made a joke about poetic time then apologized and offered to buy me lunch. When the semester was over and I was done grading his assignments, we'd become friends or acquaintances or something in between. We'd chat outside some days, walking along the brick sidewalks of our small college on our way to classes. A few times we got lunch, and often Stephen and I—and David, our mutual friend—would cross paths near the hydrangea-lined entrance to the library, where we seemed to find ourselves at all hours, though we rarely left with any books.

I don't remember how Wittgenstein came up that night on the train, only that both Stephen and David had heard of Skjolden—the village I'd thought so obscure—and suggested we meet there to make a pilgrimage to Wittgenstein's cabin. It was less than a month until all three of us moved a quarter-turn around the globe; I was going to Norway and both guys were leaving for separate moves to Germany. David was going to study Jewish history and the Holocaust; Stephen was beginning a four-year PhD in philosophy; he needed a change of pace after a hard winter; he wanted to study European philosophers in Europe, *in their home countries and hometowns—in the places that they lived—it only makes sense.* I'd wanted to get back to Norway—the mountains, the guesthouse, the alpine air—ever since I'd left.

That night, Stephen wrote his new number on the back of a receipt and then slid it into my coat pocket as we found adjoining seats on the dimly lit train. Below his number, the note said, "July? August?"

My first year in Skjolden, I didn't know about Wittgenstein's cabin, or even that he had lived in the village. I'd had little time to learn Skjolden's history; when we weren't hiking mountains that summer, we were working on the house, which was a mess of unfinished projects. When Darrell and Annette had bought the place, several of the back windows were cracked, the white wood siding was gray with age, and most of the rooms inside the house were filled with junk: broken chairs, odd ski poles, bed frames, torn curtains, lawn equipment, and old kitchen appliances. You could barely make it through the front door.

That first summer, Annie, a few other friends, and I helped Darrell tear out the attic. Termites had eaten out most of the roof beams and all of the attic's wooden walls by the time Darrell and Annette moved in, so that summer we took crowbars to the walls and ceilings, threw all of the rotting wood and insulation out the third-story window into a metal dumpster, and then nailed new termite-treated wooden braces onto all the old existing beams so the house could be fumigated and later lived in without the roof coming down. This year the others had stayed home, but I'd come back and we were moving on to the next job: leveling out the basement floor, lowering it by twenty centimeters to create a drying room, a sauna, and, eventually, an apartment for rent.

Along with the work on the house, my second summer I was determined to find Wittgenstein's cabin. I'd promised myself that before I ventured out, to the glacier or the high mountains or anywhere else, I'd start with seeing the place where the philosopher had once lived.

Since I'd arrived in Norway that second summer, I had already attempted the hike twice. Both times I had gotten lost; once I missed the path altogether and found myself deep in the woods, covered by biting flies but otherwise alone. The other time, I started toward Wittgenstein's cabin but somewhere veered off the overgrown path to his place and ended up on another path, which led to a different house. That house was square and gray, with the roof caved in, and it was full of undersized hardback books, half charred. I took some photographs; I touched the crumbling pages; I wondered if these were the same pages Wittgenstein had penned, had traced with his own hands almost one hundred years before, but then I looked up and noticed a 1960s-style plastic electrical wiring box hanging from the ceiling.

From the nearby highway, I could clearly see the metal bucket hoist that Wittgenstein had used as a pulley to retrieve groceries and other supplies from the lake. But when I got onto the ground, in the woods, I always got turned around. When I finally asked someone in the village for a map, what they gave me was a copy of a rough drawing Wittgenstein had penned himself. A simple, pencil-drawn diagram, the north/south/east/west axis was turned ninety degrees. The mountains, lake, river, and fjord were labeled as was a small black dot that said "House of Wittgenstein," but no trails or paths were added; the map was drawn from too great a height. In the end, I took the copy of the map with me and, some months later, drew my own map over it: Wittgenstein's Skjolden set against my own.

It had been well over an hour since I'd set out from the house, venturing one after another breaks in the field, enough that I started to wonder whether I was actually walking in circles, trying one path and then not realizing I was returning, each time, to that very same path again. I was still committed to locating the cabin but beginning to wonder if that commitment should also include going back to the house and asking for clearer directions.

I paused for a moment and looked out, over the marshy landscape in front of me; some of the tree trunks were flooded in ankle-deep water, others in knee-deep. Beyond the water, the woods broke steeply upward. Suddenly there was something I hadn't seen in view before: a piece of bare wood. I moved slightly to see if I could get a better view of what that wood was connected to. Visible only from a single angle, far from where I'd started, was a glint of red: bright, almost unmistakable, the color of maple leaves in fall, of cherries straight off the stem.

Though I was tired and covered in mud well past my ankles, I felt a sudden surge of energy when I saw that flash of red. As I pushed some brush aside and moved closer, I could see it through the trees: a faded "W" painted on the side of an old shack, probably an outbuilding, maybe belonging to Wittgenstein himself. The shack, aged and missing several boards, was tucked behind a small crop of trees, trees that had hidden it from sight. Even the "W" was partially gone; its thick and crooked lines were missing several strips of red. At least it was clear, however, that I was finally on the right track.

My shoes and pants already soaked through, I decided to make a run for it, straight through the high grasses and the water. I plunged one foot and then the other into the dirty water in front of me and bolted toward the red mark. As my feet sloshed through the water, I kept my eye firmly on the red mark, never letting it out my sight, until I found myself closer and closer to it.

When I finally reached the decrepit wooden building with the "W" painted on its side, I stopped, caught my breath, and looked behind me; there was no path from the tractor trail to here, just marshy water and mountains and woods in the distance; obviously no one had visited Wittgenstein's land in months.

After his first year in Skjolden, Wittgenstein left Norway. He returned to the cabin several times, but he never seemed to stay there for more than a few months. He tried teaching school in rural Austria; working in a hospital, as a shipman, and in the Austro-Hungarian army as a soldier during the war; and even returning to the university in Cambridge. All the while, he lived in Norway intermittently, seeking to make sense of the poor reception of his ideas on language, even by some of his friends and colleagues.

In 1926 Wittgenstein's sister Gretl commissioned architect Paul Engelmann to build her a townhouse in Vienna, but then, in an effort to stave off his recurring depression and to give him something useful to do with his time, she asked Wittgenstein to oversee the project, helping out where needed.

Wittgenstein was a tyrant, the project becoming an obsession: he wanted not only to build a house but also to create art, art that took into account systems and models of logic. The house could be, it seemed, a microcosm of the philosopher's aesthetic, challenging contemporary structures of belief throughout Vienna, a study in contrasts beside the city's ornamental artistic style and dense, tree-lined streets.

The builders and workers couldn't stand Wittgenstein; even his longtime friend Engelmann eventually turned over his plans to let Wittgenstein finish the house how he wanted. When the house was nearly done, Wittgenstein demanded the ceilings be raised thirty millimeters so they were the exact proportions he had called for. Instead of curtains, he installed sheets of metal that raised and lowered from the floor on a pulley. The house was stark and white, a series of three adjoining blocks, which Wittgenstein himself would later describe as "cold." In the end, Gretl would decide not to live there.

Some people said, even then, that the house reflected a change in Wittgenstein's philosophical beliefs: a midpoint, as one scholar put it, "a hinge." The cabin was different; it was simpler, more idyllic: the place Wittgenstein always said he'd settle, though he never quite could.

It took weeks to clear the guesthouse basement floor even though I threw myself into the project, so exhausted at the end of each work day I hardly wanted to walk around town or cook or swim or play soccer with the local kids.

The original plan had been to take one day and jackhammer the entire cement floor into smaller pieces of rock. The second day, Darrell, other friends, and I would shovel that rock into wheelbarrows that we would take outside and dump into the thirteen-hundred-meter-deep fjord. On the third day, we'd clean everything up and begin pouring cement for the new, lower floor. As it turned out, the jackhammering took several days and the pieces of rock needed to be cleared continually. It was heavy lifting; the work was slow and dark and loud: jackhammer a small area, chisel the cement in that section farther, by hand, then load the larger pieces of rock into the wheelbarrow, walk all the way out to the fjord, dump the wheelbarrow, and repeat the process.

Once the whole floor was jackhammered, the foundation was a mess of uneven rubble and pieces of white rock, so fine we had to use brooms and dustpans, not shovels, to load the piles into wheelbarrows to take outside. For hours, I swept rocks into dustpans and then listened to those same tiny rocks clank as I emptied each dustpan-full into white two-gallon buckets or the wheelbarrow, if someone else wasn't using it.

A thin layer of dust covered everything in the house: the furniture, the walls, the fixtures, our clothes. Even when someone finally thought to shut the basement door, the dust came in through the cracks in the doorposts and the window frames and the slats in the floor. When I thought about that summer in the guesthouse, some months later, that's what I remembered most: how that layer of white attached to everything; as much as we cleaned, we couldn't get rid of it. Even when the basement was done—leveled out, new cement poured and polished—a concrete fog still hung in the air, never quite falling, suspended for days.

However much time and energy it took, the work still made sense to me. While I lived in Chicago, I had stumbled into a part-time job of fixing up old houses: tiling, painting, sealing concrete, hanging lights, and refinishing floors. Though I didn't have any construction training or background, I liked working with my hands and the challenge of figuring out problems in real space and time.

The first room I ever painted was at the house of a professor, Anna. The house was a hundred-year-old Victorian, brick and sprawling. That first day, when I walked inside, the rooms were dark and the wood subfloors were bare, with oriental rugs thrown over sections of dried carpet glue and nails. One bathroom was pink, countertops and all; the other had lime-green, plastic-tiled walls and ceiling. The dining room that I was hired to paint had walls that were two different tones of mustard, divided by a chair rail, and a dingy off-white wooden ceiling. By the time I had finished painting, later that first afternoon, the room was a soft, almost flaxen yellow, and the trim and the ceiling were such a bright white that the room looked like it might lift off from the rest of the house.

After Anna's house came an offer for painting from one of her friends and then from someone else at the college. Soon one of my coworkers from my part-time tutoring job asked if I'd consider helping her redo the shingle siding on her garage. After that job, someone else asked if I knew—or could figure out—how to take out old carpets and install baseboard and trim.

When I wasn't fixing up old houses, I was thinking about them: paint colors that would look better on the walls, carpet that needed to be stretched, what I would do with the siding or the front porch if I were the owner. Sometimes I fixed things the owners didn't know were broken: a leaky faucet or a crack in the plaster. From my own white-walled rental apartment, I searched real estate listings for houses I could never buy, made plans for their renovation on napkins and spare corners of paper. I even fixated on the guesthouse, my imagined endings for its various rooms seamless, easy.

After that first red "W" on the side of the shack, I began regularly seeing small red o's—like bull's-eyes—painted on trees lining the overgrown trail in front of me. I followed the bull's-eyes up for some time, through the thick brush-covered woods that opened onto the lake, only once missing a bull's-eye and ending up at a sheer stone wall, but I backtracked and soon was again on the trail of rocks beside high evergreen trees along the ridge.

The red markers began to come closer together as I continued along the trail. I hurried from one marked tree to the next as fast as my feet could move. I sprang over mossy rocks and through a wall of dangling evergreens, past streaks of light breaking through the dark needles and onto the low mountain trail. I walked through more than one spiderweb and rapped my legs and arms continually to try to beat the mosquitoes and flies. It wasn't long, though; perhaps a mere fifteen or twenty minutes after the "W" on that shack, I clambered up a last rocky hill and, finally, was at the top of the trail.

In the small clearing up there, the air was sharp and warm. To one side was a steep drop-off to the lake; on the other was dense forest. In the middle of that clearing, in the place where all the photographs and maps had shown a cabin—roof neatly tiled, with white trimmed windows, the thing Wittgenstein had built—there was a rock foundation surrounded by trees. It was low and square, a few makeshift steps leading down, then a layer of leveled-off, unevenly spaced rocks, a meter or so high with a small rectangular hole in the center, probably a root cellar. Much of the foundation was covered in moss, and grasses and small plants were beginning to overtake the rocks. A single tree—a story high—rose up out of the cellar.

I turned around and looked back, wondering if I was at another relic of an outbuilding, if I had missed a bend in the trail or needed to continue on ahead. Then I saw it, lying across one corner of the foundation: a small Austrian flag, red-and-white striped with a faded wooden pole. I was at the end of the trail, but the cabin was gone.


In the mid-1930s Wittgenstein began to recant his early *Tractatus* philosophy on language, describing it as "overly narrow" and "lacking context." In his early years Wittgenstein hadn't accounted for multiple meanings of words and for the way, whatever our context is, we try to make word meanings appropriate to it: languages. Talking about language without recognizing context,

Wittgenstein said, was like "trying to walk on frictionless ice." We must, he urged readers, return to the "rough ground of ordinary language, language in use."

Instead of words acting only as object names and the combinations of names making sentences and ideas, Wittgenstein began to describe language as a game, where all the players knew the rules and made moves accordingly, each word flexing to accommodate the game that was being played and the move that was being made. His most famous example of a language game was what he termed "builder's language." "A is building with building-stones," he wrote, "there are blocks, pillars, slabs and beams. B has to pass the stones in the order which A needs them. For this purpose they use a language consisting of the words 'block,' 'pillar,' 'slab,' 'beam.' A calls them out. B brings the stone which he has learnt to bring at such-and-such a call." "Speaking a language," he said, "is part of an activity, of a form of life."

This "builder's language" was created about the same time that Wittgenstein decided to take up a new formal occupation of architecture. Though he never built another house after that one in Vienna, architecture made sense for Wittgenstein, more sense than a career in philosophy or language. It was easier to build houses that he never had to live in.

It had been months since Stephen, David, and I had all seen one another, when we rode on that train together in Chicago and they told me that a philosopher named Ludwig Wittgenstein had lived in the same small village where I was moving, not far from continental Europe's largest glacier. Even then, our meeting was just chance; we happened to walk to the same station at the same time;

we happened to be heading in the same direction, getting on and off at the very same stop. We happened to all decide to spend that rainy evening out on the town, rather than staying warm and dry in each of our own apartments.

We would not make it to Wittgenstein's cabin, or not all of us at least, and I wonder now if I already knew this, even at the time. What I do know is that night on the train we talked about the future like it was certain, like we believed, all evidence notwithstanding, that in a new place, in different air, everything that was lost could be fixed, recovered. Perhaps we thought we could find something in these places that would push up against loss and our leaving; perhaps we didn't know you can't always go back out the same door you once walked in. Whatever the reason, in the quiet heat of that passenger train, we dreamed of building something solid, something thick and hopeful, with walls and doors and windows.

Both guys were wearing shirts and ties that night on the train in Chicago, collars unbuttoned; I was in a short, gray dress. I remember now noticing how much older Stephen and David looked than they had only months before and, also, how the yellow lights of the city reflected off their faces at every stop, then stayed, just for a moment.

I walked over to it, to the remnant of Wittgenstein's cabin. There were some fresh animal tracks imprinted in the wet dirt, but otherwise only my own footsteps. I stepped carefully across the neatly piled rocks, picked up a brick, and traced my finger along its edge. I took a couple of photographs to show David and then set down my backpack and sat for a while, on the edge of a large stone, looking out at the Eidsvatnet Lake, the Sognefjord in the distance, and some small waterfalls and a tunnel lining the blacktop road a few miles away.

I tried to imagine Wittgenstein pacing or writing or yelling at the local kids to stay away, but I couldn't see him, not there, not waiting by the dead tree or leaning against the metal pole—short, thin, tousled brown hair, rounded nose—not standing on that overgrown foundation. As hard as I tried, I couldn't imagine him without his cabin. In the pictures I'd seen of Wittgenstein, and even on a hand-drawn map, there had always been that cabin: two-story, square, neatly tiled roof, white-trimmed windows. I hadn't considered that by the time I got there, the cabin might be long gone.

After a while the clouds began to appear in the distance again, and so I picked up my things to head back down the trail. I stopped, though, just for a minute, near the edge of the clearing. No more than a few meters from the house, there was a break in the trees; the center of the lake was still, but I could see farther now and realized that the water at its edges was not. From the east, it was fed by a river of melting ice; to the west, it rushed through a narrow channel, past the guesthouse and the village, and into the fjord.

It was the same water that would travel down that fjord, past more green mountains and small villages and larger towns, past car ferries and bridges and boats, and then to the very edge of the land itself, splitting into the North Sea and the Norwegian Sea and maybe eventually finding its way to the Atlantic Ocean. As I stood in the place Wittgenstein probably once stood, I wondered if what he wanted in those years in Norway was an edge, something to cut back against, some way to name all that had already happened and all that was to come.

I imagine that at some other point in time all that water might have looked like a beacon.

For a few summers after the guesthouse apartment in the basement had nearly been finished—bathroom and kitchen tiled, walls plastered, electrical and plumbing handled—Darrell and Annette stopped working on the house. Darrell was going to graduate school and Annette was watching their now three small kids. They'd enlisted the finances of other investors too, which meant any renovations had to pass by three or four sets of hands.

When I came then, instead of sawing or nailing or ripping off siding, there were frantic bursts of cleaning spearheaded by Annette: mopping wooden floors, shaking out thirty-year-old carpets, wiping down walls, making rounds through the house every day, at two or three p.m., picking up after children. There was a list of still-unfinished projects—rooms that needed painting, floors to refinish, a balcony that was hanging precariously off the side of the house, held up by a single unanchored two-by-four—but those were relegated to someday in the future, when there would be extra time or funds. The black roof got more and more chipped; the white paint, which we'd once redone, grayed against the river's spray.

I should have realized that for every project there is either an end or a point of abandon, should have reminded myself that it wasn't my house to fix anyway, that none of the houses were my projects. I didn't; instead I swept the bare floors in the attic, cleaned the kitchen countertops, and helped my friends move couches down three flights of stairs and then back up them again. All along—even though I might not have been able to articulate it then—I fell further and further into that place, like I had always lived there.

# DRIVING WYOMING

It's five a.m., still dark, and the steam is rising off the cracked blacktop centerline like fog. I would like to be dreaming, but I am not; I'm in my car, driving south from Wyoming to Colorado on Highway 287, one of the most terrifying and beautiful highways in America. The seventy-two-mile stretch I once drove every day drops three thousand feet through the Rocky Mountains, from a high plain—the Laramie Ridge to the east, the Snowy Range to the west, climbing blue-gray from the snowfields—to a low forest of red-granite boulders, ponderosas, and scotch pines.

As often happens on 287, I see no one: no cars, no semis, no bicycles. The landscape of the road is stripped down, bare-boned. Few people make this drive; 287 is remote. It's two-lane and it's cut into the land in a series of narrow canyons and sharp mountain passes. In the hour-and-a-half drive, there's one sometimes-open flea market/post office, two closed cafés, three churches, and thousands of acres of dry ranchland.

I roll down my driver-side window. For the short space of the drive, everything is immediate: the empty road, the air that smells of brush and pine, the long spaces of quiet that it's easy to mistake for calm in a cold, windswept spring in the Wyoming West.

On April 29, 2009, the news began to break and then swell, first by email then phone then radio, then online literary journals, local television stations, and national newspapers, that forty-two-year-old American poet Craig Arnold—author of *Shells*, author of *Made Flesh*, winner of the Rome Prize, the Yale Series of Younger Poets—had disappeared two days before while on a hike on a volcano in Japan.

Craig was on a research trip in Asia. He'd taken a ferry from a nearby Japanese island, those two days before, checked in at a local inn, drunk a cup of tea, dropped off some bags, and gotten a ride to the sandy base of an ancient strato-volcano, Mount Shintake, from a village resident. He was wearing dark pants and a nylon jacket, carrying a phone and walking sticks but no food or water. The volcano hike was supposed to be a quick hike, an easy hike, rated for difficulty two out of ten. He planned to be back to the inn for dinner, the innkeeper and local resident said.

He wasn't back for dinner. By nightfall Craig hadn't called his family; he hadn't returned to the inn. He hadn't been seen by the guests he'd befriended on the ferry to the island, the resident who gave him the ride up the mountain, or any of the people who lived in the neighborhood houses near the mountain road down. Instead, it seemed that he had disappeared, vanished, maybe got lost—no one knew—on a four-by-twelve-kilometer island of 160 residents and three volcanoes, in a country the size of Montana but containing 124 volcanoes, in the Ring of Fire, a region less than .01 percent of the world's landmass but with 452 volcanoes, three-quarters of the world's active volcanoes.

While the rest of us were sitting around long seminar tables discussing poetry or fiction or pedagogy, writing in chalk on the blackboards in our classrooms, walking, running, driving around town, out of town, through town, Craig was hiking volcanoes, smoking mountains imbued with the sublime, with the *perpetual presence of the sublime*, making space for the sublime if the sublime is not an emotion but a meeting, a seeking. "He was going," his sister-in-law would later say, "where ordinary people can't or won't go, to tell what that experience is like."

I slow down, turn off my headlights, and let the dark settle through my car until I can make out the lines of the road by the way the blackness changes—an eighth of an inch in gradient at most—the difference between an unyielding dark and a dark someone might find their way in.

I first met Craig in spring 2007. He was teaching poetry at the University of Wyoming. I was visiting Laramie from Chicago, thinking about leaving my temporary job to go back to school but unsure whether I wanted to move to a city that was so small, had so much open air. It was a chance meeting; I got the times wrong and showed up early for a different class; Craig walked into the otherwise-empty room. I don't remember what Craig told me about the university or the landscape or if he said anything about these at all, only that he was wearing a buttoned-up, blue-striped shirt and black leather jacket when I met him, that we were in a small library, that his forehead gleamed in the light like an orange or a glass ball or something rough that had been polished smooth.

As it turns out, I didn't move to Wyoming; I moved to Colorado and began commuting between the two states. While most of my friends were giving up their cars for city jobs, closet-sized apartments, and grocery stores that stocked no fresh fruit or vegetables but were in Park Slope or Brooklyn Heights—anywhere there was the thrill of buildings, bodies, the subway of another city to conquer—I went back to school, began running in the foothills in the early mornings, and spending three hours a day on the highway.

I told people as an explanation, "My cousins live in Colorado," which they did, but this wasn't why I moved to Fort Collins instead of Laramie or why, twice daily, I chose to drive 287 instead of some other route. At the time, I couldn't lay my finger on exactly why I did this beyond that I felt unable to settle into the life of any particular city and strangely compelled by the driving: the never remaining in one place, the constant motion along the edges of somewhere so vast as to be almost imperceptible.

In his travel writing class, six months later, Craig asked everyone to come to the semester with a writing project in mind, something we'd start a short collection around. Though I considered writing about Chicago or the glacial landscape I had returned from that fall, in the end I chose 287. Nothing I handed in that semester ended up concerning the highway. As many times as I tried to write about the road, I never quite could.

In ancient myth, Japan's great throng of volcanoes was birthed when the god Izanagi, the father of the Japanese islands, beheaded his son Kagu-tsuchi, the deity of fire, and chopped his body into eight pieces. The act was a retribution for Kagu-tsuchi's prenatal crime of burning his mother, Izanami, to death during his childbirth. The eight pieces of Kagu-tsuchi's body became eight volcanoes. Although Izanagi's tears and Izanami's belly continued to birth gods and goddesses as Izanami lay dying, with Kagu-tsuchi's birth their co-creation was over. The rise of the volcano—in Japanese, *kazan* or "fire mountain"—signaled, in effect, the beginning of an end.

The geographical impetus for this mythology is clear: Japan, itself, sits on the intersection of four tectonic plates—the Pacific, Philippine, Eurasian, and North American—and as a result, the country contains 10 percent of the world's volcanoes. The Global Volcanism Program, a project of the Smithsonian that "seeks better understanding of all volcanoes through documenting their eruptions, small as well as large, during the past 10,000 years," has mapped all the known volcanoes in the world as red triangles on a world relief map. In Japan, the triangles fork, heading northeast, northwest, and southeast in almost continuous lines, each triangle pressing up against another.

This was part of the reason Craig went to Japan: this, to climb Mount Fuji and to follow in his poet-mentor Bashō's literal steps. Craig was writing a collection of prose poetry on volcanoes, a project he planned to call *An Exchange for Fire*. He'd already summited volcanoes in Peru, Nicaragua, Greece, Colombia, and Guatemala, all before Japan. Japan was Craig's final trip before completing the book.

That midafternoon in late April, Craig did what he'd been doing for months and years; he set out on his own for an easy hike, probably taking mental notes and maybe photographs as he went, anything that could help him make poetry. This was simply one more in a long string of volcanoes for Craig, a hike that should have taken three, maybe four, hours.

*Cold and windy or dark and pathless*, writes Craig, in an online travelogue of his journeys through Japan, *what is this forest in which we find ourselves? Or rather where we lose ourselves to find our way out? A destination needs desire. To reach it requires will. The wanderer has will without desire, to move without getting anywhere, but to keep mov-*

ing... *it is like the shark who must keep moving, moving to breathe, moving to stay afloat, or else sink, into the dark blue depths, under the weight of endless tons of water, where even the light of the sun, if it could reach that far down, would be pale and cold.*

*One need not shrink from the sublime,* Craig writes in another entry on the same site, in response to Wallace Stevens's poem "Esthétique du Mal." *Nay, one may rather seek it out, with a pack on your back and a stick in your hand, liberal applications of sunblock and when necessary a gas mask over your face.*

Twenty miles outside Laramie, I drive past the backside of the brown "Welcome to Wyoming" sign. On 287 you feel a difference between one place and another. Unlike so many geographies, here the lines are mapped right. The Wyoming stretch is open, crossing a high plain and an even higher ridge with miles of mountains on either side. As you reach the Colorado border, suddenly there are trees, evergreens: lodgepoles, ponderosas, and Scotch pines with clusters of green-blue needles pressed up against and falling over the road. There are rocks too—red granite boulders—bigger than my car, rising out of the hillside, balanced between trees or under them, or perched, on their own, off to the side.

A few minutes after the Wyoming sign, through the line of evergreens, I descend one long hill and up another into Virginia Dale, Colorado. The hills are it for the geography of Virginia Dale—then one dirt road, a couple of houses, and a church. At the bottom of the hill, there's a bronze plaque reading, THIS MEMORIAL IS PROPERTY OF THE STATE OF COLORADO and *Famous Stage Station on the Overland Route to California, 1862–1867... Vice President Colfax and party were detained here by Indian Raids.*

At the crest of the hill is the boarded-up post office, peeling peach-colored paint and a half-green, half-tin roof; there's a white mobile home just behind it—also empty—and a large gravel parking lot in front of it where truckers sometimes park for the night in snow or rain or ice.

"He's lost and he needs my help." This is what Craig's brother, Chris, said. "My brother doesn't have a great sense of direction and uses a GPS to find my house in Brooklyn, but he's not a person who takes stupid chances. He's lost and he needs my help."

Everyone, it seemed, believed Craig needed their help. When Craig didn't return to the Watanabe—the inn where he was staying—by eight then nine p.m., the innkeeper began to worry and contacted the island's fire brigade to warn them a foreign hiker might be missing on the island. Within a few hours, volunteers had driven the roads along the bottom of the volcano and climbed all four of Mount Shintake's well-marked paths looking for Craig and calling his name.

By the following morning forty people were searching the area; a team of policemen came over from a larger island to coordinate the investigation and contact American authorities and Craig's partner and son. Rescuers and locals went into the densely wooded forest on foot with search dogs looking for alternative routes or shallow recessions where Craig might have slipped and been injured. Telephone and satellite companies tried to make contact with the GPS on Craig's phone; a military helicopter was employed to circle the volcano and the island's coastline.

Back in Wyoming, our department chairs, Beth and Peter, and professors, Brad, Alyson, Harvey, and Kate, fielded calls from the media, talked to Craig's family and friends, and sent emails to the graduate students and English department, trying to piece together as many details of the situation as possible. My office mates, Dixie and Tyler, and I gave each other updates when we drifted in and out of our conference-room-turned-office between teaching and our own literature or writing classes: footsteps discovered that might have been his, updated times and sightings of Craig by travelers and locals earlier in that trip and even that day. Mostly we talked about how it didn't seem real, that surely, in a few weeks' or months' time, Craig would be back in his office, just upstairs.

We were nearing the end of the term when Craig went missing; the late-spring high-altitude sun was already streaming through the windows onto the oriental rug, the long wooden table in the corner, and the paintings on the wall.

When I was a child, I had a book called *Pompeii…Buried Alive!* It was the story of the sudden volcanic eruption of Italy's Mount Vesuvius and the later archaeological find of the city of Pompeii. *Once there was a town named Pompeii. Near the town there was a mountain named Vesuvius. The people in Pompeii liked living by the mountain. It was a good place to grow grapes. It was a good place to raise sheep. And—it looked so peaceful!*

In the book, there was a red tree of fire under the grassy mountain, wide at the base, like a ball, then shooting straight up through stacked layers of purple and gray rock until the top layer of rock, when the rock became smooth and the red shoot became branches, all curving upward to the open air like a maple tree, or an oak, in spring.

*The day started out the way it always did,* the book continued after that picture of the fiery tree. *The sun rose. People began coming to Pompeii with things to sell. Fishermen were bringing fish. Peddlers were bringing melons and straw hats. Farmers were bringing vegetables. Shepherds were bringing sheep. Carts rumbled through the narrow gates and into town.*

I didn't remember that book until the week Craig disappeared, but suddenly I began to think of it often. I asked my dad if he knew where it was in our house. After he looked through old boxes of books in the basement and couldn't find it, I tracked down a copy online from a local library. It seemed strangely important, somehow, to return to where I first learned about volcanoes, to see again that image of the tree under the mountain.

Road conditions on 287 are sometimes bad and, more often, unpredictable. *Whiteout, blowing and drifting snow, rain, fog, strong winds, ice;* these are the road condition categories on the Wyoming Department of Transportation website. Each of them is cycled through regularly, sometimes in just a few days. When things get particularly dire—drifted and blowing snow, visibility less than a quarter mile—the road closes altogether, stranding travelers in Fort Collins, Laramie, or sometimes—like those trucks on Virginia Dale hill—between the two.

Even when the road's at its best, it's still dangerous: at least forty people have died on the thirty-five-mile Colorado stretch of 287 in just the past ten years. I once saw a map of these deaths in the newspaper. In that newspaper article, a red dot was placed over each mile marker where a fatality had occurred. At mile 386—Ted's place, named after 1920s Colorado senator Edward "Ted" Herring, and the spot where 287 merges with Colorado 14, an east–west highway along the continental divide—there were eleven red dots. At mile 380, five miles short of the Wyoming border, five more. At marker 360, Owl Canyon—one of the few places in Colorado where alabaster is found—there were eight dots, at mile 353—just after 287 leaves Laporte—there were six.

In 2001, on the Wyoming side of 287, a crash killed eight University of Wyoming student athletes, all cross-country and track runners. Seventeen miles north of Laramie, their SUV was hit by the pickup of another student athlete from the rodeo team at 1:30 in the morning on a Sunday in September. After the crash the Wyoming Department of Transportation performed a statistical analysis of 287. The study found the probability of crashing on 287 to be similar to other Wyoming roads, but crashes on 287 were twice as likely to be fatal.

Six months into my own driving of 287, a man on a cell phone several cars in front of me misses a turn and steers his car over a steep bank. Six months after that, the truck of two university nursing professors spins out, sending another car over the edge, killing both the professors, injuring four others. And six months after that, my friends Matt and Adam flip their truck leaving 287 for a dirt road. Matt is fine; Adam—a builder—breaks his arm. Twice, my own car dies on 287; once I blow out a tire on the inside of a blind curve, but I don't realize it until I hear my metal tire rim scraping along the asphalt.

Still, some nights when the weather is clear, I drive while watching the sun fall over the mountains and the wood-post wire fences by Cherokee Park Road or the dry streambed just below it. Occasionally trucks fly by, but more often than not, except for the steady hum of my moving car, everything is quiet. The long grasses, the red dirt, the farm fields: in the early evening haze, even the dusty highway is streaked with a kind of light.

Craig's classes met often at bars or restaurants or his apartment instead of a classroom: anywhere there was alcohol and baba ghanoush. Craig was all energy, frantic in his teaching and in his critiques of the poetry that we wrote. *More pressing*, he scrawled sideways, across one of my pages, right over the text, *this is it! finally* on another.

We were reading Bashō in Craig's travel writing course before he left. *The moon is brighter since the barn burned. The temple bell stops but I still hear the sound coming out of the flowers.*

> should I take it in my hand
>     it would melt these hot tears:
>         autumn frost
>
>     dew trickles down
>         in it I would try to wash away
> the dust of the floating world
>
>     how easily it rose
>         and now it hesitates
>             the moon in the clouds

We talked about many things: Wyoming, world literature, food, our travels and attempts at travel writing. One student was following pronghorn through Yellowstone that spring, tracking their numbers and mapping their shrinking migration corridors; another had come back from Guatemala after two years in the Peace Corps working as a cowboy during Jueves Negro; someone else had biked across Greece, visiting ruins and sleeping in temples; Craig was hiking volcanoes; I was driving.

Still, though the class was full of talented poets and writers, I remember clearly only a single line from any of the other students' work. *There was silence,* the woman wrote about a phone conversation with her ex-boyfriend, *and the sound of something breaking.* We were sitting at a long wooden table when she read this, right next door to the library where I'd first met Craig. The lights were a blunted yellow and it was snowing outside, hard—this I remember—so hard that when the highway opened two days later the endless drifts had washed clean the plains.

Halfway through my drive along 287 I see the Virginia Dale Church, a white clapboard chapel that seats thirty with a cemetery and two outhouses and only meets on the second Sunday of each month. There are two churches along 287 and an abbey, though I don't discover the second church—the Livermore Community Church, a broad tan and brown-trimmed building that backs to 287, a ways off the road, so you can only see the steeple from the highway—until a few weeks before I leave for good.

The St. Walburga Abbey, founded on Psalm 23 and the other psalms of trust, is the largest of the three churches and simultaneously part of and separate from the landscape. It's gray and stucco with a bronzed roof, and it's built right into the edge of the rocks. Or this is what the abbey's website photographs show; it's set away from the road so I never actually see it. Like this day, all I ever see is a small metal sign on the edge of a dirt drive leading to the abbey, right before the top of the Virginia Dale hill. I sometimes slow down, but I always end up driving by.

One day, I almost stopped and even turned onto the dirt drive. There was only a light snow that afternoon when I left Laramie;

people were walking and biking around town. The blowing snow started in Tie Siding, eight miles north of the Colorado border. By the time I reached the Virginia Dale hill, I couldn't see the road, only wide swaths of white, gusting and then piling into drifts. I thought about stopping at the abbey that day and even practiced what I might say to the nuns who lived there if I went inside. *The body is a unit*, I remembered reciting in the church of my childhood, *for we were all baptized by one Spirit into one body—whether Jews or Greeks, slave or free—and we were all given the one Spirit to drink.*

In the end, I drove only a short ways down the road to the abbey before changing my mind, turning around, and deciding to keep moving.

Within the first week of his disappearance, the Japanese government and the U.S. Air Force joined in the search for Craig. Soon a private NGO, 1st Special Response Group, went to Japan to look for Craig. Fulbright volunteers, too, came to the island to walk in slow lines up the mountain and canvass off-trail areas that hadn't been combed through or couldn't be seen by helicopter or airplane. A fund was started and a Facebook page and a blog, all called "Find Craig Arnold." People around the world—many who did not know Craig, some who did not even know his work—were donating money or writing their representatives and senators in hopes that the U.S. and Japanese governments would intervene or that the private search group might have the funds to look harder and longer.

At first, the investigation seemed to pay off: footprints that looked like Craig's were found heading up one of the volcano's trails but not coming down. Japanese officials posited that this meant he made it up the path but got disoriented near the summit and took a different path or what looked like a path but wasn't a path down.

Still, as the days went on, no one could quite tell where the obscured footprints led and whether Craig had actually summited the caldera of the volcano and then walked over to another stratovolcano site, or if he had tried to come straight down the mountain and slipped somewhere along the way. *Craig's tough; he's an avid hiker, he'll be fine*, faculty and students in Wyoming told one another in the long hallways of our old English building and out on the browned-grass quad after each new piece of seeming evidence was relayed and then refuted, *it's only a matter of time*.

We didn't know if it was a matter of time, or how much time. All we did know was that Craig was still lost and the Japanese island that he'd been hiking was dense; it was filled with trees and grasses. It was nothing like Wyoming there, where Craig was: it was green; it was swelling.

Craig spent a couple of days in Wyoming, shortly before he left for Japan. Though he'd been traveling on leave already for some months, he had decided to return to Laramie to give a reading from his latest book and from his new volcano prose poetry.

It felt as if he'd swept into town that visit with just enough time for a few meals and to repack his things for the apocalyptic landscapes ahead. The posters advertising Craig's reading appeared hastily made too; they showed a dark granite cliff—probably some ocean edge—with steam and water pouring over it. Just left of center, though, two electric-orange shoots of lava—like a cartoon or a piece of clip art—were superimposed on the cliff, pouring right into but never mixing with the water.

Neither Dixie nor I could make Craig's reading, but we still hung that mockup of a volcano poster on our office door, next to a couple of advertisements for the MFA reading series, a list of semester dates and deadlines, and a signup sheet for student conferences. Some of my friends did make it to the reading and told me that Craig read with urgency and with flair, throwing his scarf into the audience at one point, beating the lines of his poems like a snare drum on the bookshop's wooden podium.

Craig told a story, too, on that visit, of his encounter with a small bird on a volcano in South America. Craig and a fellow hiker and guide were coming up a trail. The air was filled with smoke and ash, so much that Craig covered his mouth and nose with a scarf, his head with a hat. Suddenly, there as they rounded a bend was a bird, tropical and hopping in long fluid strides, right along the path, seemingly unaware of the landscape around him. That was it: Craig relayed the story without interpretation or commentary; the bird was a poetic observation, a fact, just another thing that seemed to draw him.

Perhaps—I will think as someone tells me this story later—some things had already begun to come apart.

On April 24, 2009—three days before he plans to hike Shintake—Craig writes:

> The day is breaking—
> one side of the mountain pink
> one in cold shadow

As the days went by, the news reaching Wyoming grew increasingly bleak. Five days after Craig went missing, searchers determined his steps didn't lead to the caldera of the volcano and that he was lost elsewhere along the trail. Nine days after Craig went missing, the ISRG search-and-rescue group believed they found his trail, leading toward the edge of a ravine. Two and a half weeks after he went missing, a technical climbing team from Tokyo, the Canyons, were hired to go over the cliff, belay down to the bottom, and search for Craig's body in the thick vegetation.

*Body*: this was the first mention of a body. It made sense; a person survives only so long in the woods, without water or food, injured, unconscious, possibly taking a long fall.

"I felt like I knew Craig," people who did not know Craig were writing or calling to say, even that soon. To legitimize the loss? To brace themselves for it? Or maybe this is what we talk about—our knowing—because there cannot be a loss of what is not known. Without a loss, there's only a gap—*an empty space or interval; interruption in continuity, a wide divergence or difference, a break or opening, as in a fence, wall, or military line; breach, a deep, sloping ravine.*

Craig and I talked about 287 just one time, a few months before he went to Japan. We were in his office; I was sitting in an old green chair beside a white fur rug; he—tall, lanky, legs stretched out—was leaning on a metal desk. *Why do you do that drive?* he asked me, nodding slightly, like he knew, before the question, the answer. *287 is a narrative of life*, he said, *your life*, though I can no longer remember whether he said "narrative" or "story." I sometimes try to imagine his voice—quick, crisp, accenting the ends of words—and to listen for which phrase sounds more right in that cadence, but I cannot hear him. It seems the distinction should not matter anyway, but it feels like it does and like it might have to Craig, who titled his book *Made Flesh*, after *the word made flesh*, and in it began his "Hymn to Persephone," *Help me remember this.*

The Wyoming West, you see, was not enough for Craig: not the stable, everyday university life in a mountain town, not the wide expanse of sky. Craig lived in a barely furnished apartment—not a house—rented, not bought; he left on long weekends, in the summers, for winter breaks, went to Denver, Salt Lake City, Chicago, Manhattan, Athens, Rome, Japan, any place exciting, any place with the possibility of something more, something else. There was no illusion of permanence there in the West, not for him, not for me. There was only art—I realize now but did not then—only poetry.

Craig gave me two books that day I met him in his office. He took them right off his bookshelf and told me to read them and keep them, that they would change the way I thought about my writing. The first was *The Rings of Saturn*, a part-biographical, part-fictional record of W. G. Sebald's walking tour in eastern England but a record that begins with a stint in a hospital and Sebald's telling of the "paralyzing horror" that comes over him when "confronted with the traces of destruction, reaching far back into the past, that were evident even in that remote place."

The other book, *The Lady and the Monk*, is a story of a much different trip, a trip through Japan where Pico Iyer meets his future partner and wife and discovers he can reside happily between two places and cultures. "I am simply a fairly typical product of a movable sensibility.... I am a multinational soul on a multinational globe," Iyer writes. "Taking planes seem as natural to me as picking up the phone or going to school; I fold up myself and carry it around as if it were an overnight bag."

That spring I kept driving, even after the Canyons could not locate Craig's body, after the conversation changed from "finding Craig" to "recovering Craig" to "bringing Craig home," and then the search was finally given up altogether amidst a steady stream of toasts, memorials, and readings of Craig's work in Laramie, Denver, Salt Lake City, and New York.

One night I drove through a blizzard alone, snow falling onto my windshield in heavy clumps, wet and then so icy my windshield wipers stopped working and I had to hang my head out the car window to see. Another, I slid on the wet pavement toward the edge of a sharp bank and had to back my car up, slowly off it. Many times, I prayed for God and angels in a way that I have not since, probably never have before. *I wouldn't mind seeing them tonight*, I called to the road.

I knew that eventually I too might find myself in danger and alone. I knew this and yet I still did the drive, every day, sometimes at night, even when I left town and could hardly see the road underneath the white, white snow. Is this what Didion calls the *mechanism of terror?* That we go forth willingly, that we bury ourselves in it, tell each other that the only way around it is through it—is *to seek it out*—that all that matters is that we keep moving, that we do not stop.

"Danger has a way of cutting through melancholy," Craig writes, "the real fear blinding you to the fear dimly imagined. If you could only always just have escaped death, you would never be sad again." *But would we never be sad again? I want to ask him now. Or would we be sad always or afraid?* Might the terror of the death-just-missed ever remind us that we live in bodies which have limitations, which are faltering, which at any point might be swept up as easily as the headlights in the distance in the early-morning sun? *I never saw Craig before*, I tell a friend who knew him like I did, *but now I see him everywhere.*

In his final travelogue entry before he disappears—before we're left only rough outlines with which to think about his last days and life—Craig writes about a plant called ashitaba, "the tomorrow leaf." *They say it grows so quickly that leaves picked in the evening will be replaced the next morning*, he writes. *Or it may bring more tomorrows.... Crushed in the hands, the fresh leaves are sweet, slightly musky—not quite mint, not quite juniper. It is a clean, windswept smell, the smell of meadow, of England, of green, the smell of a road after rain. It is the smell of a world in which there is nothing rotten or putrid or sulfurous, a world in which all of those things have been rinsed away.*

This day there is nothing to rinse away. I've traveled and moved to Norway, back to the Midwest, and then to Norway again, but I'm in Wyoming for a visit and decide to drive 287 both ways.

As I once did every day, I've left the southern edge of the Snowy Range in my rearview mirror, and I'm driving toward the northern edge of the Front Range, moving between the place I used to live and the place where my life is now.

It's cold and quiet; the sun has risen. I'm in the last stretch of the drive, nearing Owl Canyon, the final hairpin curve before I make my way into Laporte and nearby Fort Collins, but I've stopped early, gotten out of my car, and I'm standing on the edge of the road in the uncut grass. The road is dusty, but the mountains in the distance are snowcapped. Another car ahead of me has stopped too, driver-side window rolled down, a man's arm resting along it. There's a field somewhere below the road with circular bales of hay spaced at almost-even intervals.

I've stopped driving because, at last, I want to stop but also because I must. Fifteen feet in front of me, crossing the centerline, is a herd of running pronghorn, forty or more, slender, legs outstretched, flying over the road and down the ridge just past it to the open grassy valley, two or three at a time. They're close enough I can see their reddish-brown hair and light bellies, strips of white along their necks, hair raised slightly, and moving. I can hear their hooves on the blacktop, clicking for a moment between strides.

# LOST

## An Inventory

### /LÔST, LÄST/

ADJECTIVE

1. no longer possessed or retained
2. no longer to be found
3. having gone astray or missed the way; bewildered as to place, direction, etc.
4. not used to good purpose, as opportunities, time, or labor; wasted
5. being something that someone has failed to win
6. ending or attended with defeat
7. destroyed or ruined
   —Dictionary.com

### LOST JEWELS

In 2013 approximately $300,000 worth of jewels were found in a small metal box marked "Made in India" that was halfway buried, halfway sticking out of the thick, icy crust of a glacial peak on France's Mont Blanc. The tin box was about the size of a shoebox; the jewels inside it included emeralds, rubies, and sapphires—roughly a hundred stones in total—each tucked in a small sachet.

The box of jewels was discovered by a twenty-year-old French student and mountaineer. When he found the box, he immediately carried it down the mountain to the nearest police station. The jewels could have been worth millions for all he knew, but he said he would give them to a museum if no one claimed them and he was entitled to the find. Though it was one of the most valuable mountain discoveries, the student never released his name to the public.

It turns out, though, that the jewels weren't random or lacking an owner. They had been carried across Europe—from Bombay to Delhi to Geneva en route to London—by Air India Flight 101, a flight that on January 4, 1966, crashed into the Bossons Glacier on Mont Blanc at 15,584 feet, killing the flight's crew and all its passengers. The jewels had been there ever since.

## LOST IN ICE

On September 19, 1991, Erika and Helmut Simon, German climbers visiting the Alps, were hiking Mount Finailspitze near the Austrian-Italian border when they noticed a shoulder and then a skull protruding from the ice, half in the open, half-buried, facedown. Assuming it was a recent death, Austrian officials came back four days later, chipping the body out of the ice with jackhammers and pickaxes.

There was some clothing made of skin near the body and strange tattoos on its right knee, left calf, and spine. Its forehead was partially decomposed, but it still had hair and a dagger and a copper axe. When the body was taken back to a morgue, it was dated to be 5,300 years old: Europe's oldest mummy. He was named Otzi after the Otz valley where the Simons discovered him.

In 2002 glaciologist Lonnie Thompson, hiking near the edge of a receding glacier in Peru, found a plant that was 5,200 years old, the same age as the famous iceman, Otzi. This suggests, he writes, "that the present warming and associated glacier retreat are unprecedented in some areas for at least 5,200 years."

In August 2004 a local mountain guide, Maurizio Vicenzi, found the mummified bodies of three soldiers, hanging upside down from an ice wall, near San Matteo, Italy: soldiers—it was soon decided—who had fought in the First World War. A love letter was found in those same mountains and a soldier's diary.

Only a year later Vicenzi again made a find: this time an entire wooden cabin emerged from underneath the ice. The cabin was used as a supply station in the war, and was set between a one-hundred-foot tunnel drilled into the ice and a several-thousand-foot cableway that soldiers used to bypass the glacier to get to the front lines.

In the summer of 2006, in a melting Norwegian glacier, a leather shoe was found that dated back to the Bronze Age.

In 2010 a ten-thousand-year-old hunting weapon was found in what had once been a frozen ice sheet in Yellowstone National Park in Wyoming.

In 2013, not long after the Mont Blanc jewel find, a message was found in a bottle in a cairn near a Canadian glacier. The message was left in 1959 by an American geologist named Paul Walker, who told the recipients that he had placed his note and bottle exactly 168 feet from the edge of the glacier and if they wanted to see the movement of the glacier, they could measure from the bottle to the glacier's endpoint.

The biologist who found the bottle measured; the bottle and the message were 401 feet from the nearest edge of the glacier. In just fifty years the glacier had receded more than 200 feet.

## LOST COIN, LOST SHEEP, AND LOST SON

In Luke 15, the Bible narrates three parables about lostness: the lost sheep, the lost coin, and the lost son. In the first parable, a man has one hundred sheep and loses one. In the story, he leaves the ninety-nine to go look for that one lost sheep and searches until he finds it. In the second parable, a woman has ten silver coins and loses one. In this story, she lights a lamp, moves the furniture, and sweeps until, like the man with the lost sheep, she finds her lost coin. In the third parable, a man has two sons. One stays home and works faithfully while the other sets out and squanders his inheritance, partying and living recklessly until he has no more money and finds himself living among pigs. When that lost son finally comes home, the father runs out to greet him, bringing him a ring and robe and sandals; he throws a feast in the lost son's honor.

The parables seem to grow in value, at least according to the time, but they also have a common moral: the nature of a lost thing is that it should be sought out, it should be found.

## LOST: BIOLOGY AND PSYCHOLOGY

For many years it was believed that certain people and even people groups had a better innate ability to find their way than others. Recently, though, scientists have begun to tie people's ability to find their way—or conversely, to get lost—to cognitive maps, or the brain's representation of physical spaces.

Drawing on research done on rats, Paul Dudchenko, author of

*Why People Get Lost: The Psychology and Neuroscience of Spatial Cognition*, suggests different types of neurons fire in the brain when people find themselves in different physical locations. He writes, "The combination of these different cells being active in different places of the brain began to look like a kind of neural map, a representation in the brain of different places in the environment."

In his book *Lost Person Behavior*, K. A. Hill argues that spatial orientation and cognitive maps are largely taught and learned rather than the genetically instinctual or biologically based sense of direction. "No controlled study to date," he explains, "has found reliable evidence of a human ability to sense the direction of magnetic north—or any other direction, for that matter." Hill goes on to write that people who seem to usually know where they are tend to "mentally update" their geographic position as they move in their environment. People who tend to get disoriented—lost—do not.

There's also a physiological side. When we're lost, Hill adds, there's a level of fear and a level of emotional arousal. The arousal causes the limbic system to be stimulated. In small doses, this can lead to sharper mental functioning. But when the response is too heightened, it can scatter our thoughts and make us unable to concentrate or even remember things that should be familiar.

## LOST WORDS

When I turn thirty, my parents give me a necklace that reads "Not all those who wander are lost." The saying, a quote from one of my favorite writers, is carved into a silver pendant. The pendant is small and shaped like a compass, with marks for north, south, east, and west imprinted below the words.

The quote is fitting. By the time my parents give it to me, I've lived in ten cities and many more houses, apartments, and flats, enough that I can't even remember all my past addresses.

It is fitting, but I cannot decide, as I look at that necklace, whether it is true. "All that is gold does not glitter," the rest of the poem that quotation comes from begins, "Not all those who wander are lost; / The old that is strong does not wither, / Deep roots are not reached by the frost."

## /LÔST, LÄST/

According to the *Online Etymology Dictionary*, the words "lost" and "loss" can be traced back to the Old English *losian*, which means to perish or to destroy. The similar root word *los* spread to Norway and meant there "the breaking up of an army." There was also *leus*, which influenced the German words for lost and meant "to loosen, divide, cut apart."

## LOST IN SWITZERLAND

I try to see Mont Blanc one summer. I'm going to Switzerland for a couple of weeks and plan to fly through France, take a train across the country, and visit the French Alps before I make my way to the Swiss mountains. The week before I go to buy my tickets, though, terror attacks take place in Paris, and so I end up buying a direct ticket to Zurich instead.

A couple of days before I leave for the trip, I explain to a friend that I'm sad to miss Mont Blanc. "Just a second," he says, before disappearing into his study. He emerges a minute later with a map of Switzerland. The map is colorful—bright blues and greens—with mountains and towns illustrated in their approximate locations. There's also a series of passes, roads, cable cars, and even an image of a train on the map. "There. That's the spot," my friend says, pointing to the Swiss mountain Schilthorn, just past the town of Lauterbrunnen and above the village of Mürren. "You can see Mont Blanc from that pass in Switzerland."

I look at the spot where he's pointing. Mont Blanc is labeled in tiny letters on the map, surrounded by a deep blue sky.

Five or six days into my time in Switzerland, my friends and I make it to Schilthorn. They decide to hike near Mürren, but I want to see Mont Blanc—if only from a distance—and so I take a series of cable cars up the mountain. The first cars are packed—standing room only—but more and more people get off at each stop, until it's only me and a small group of Australian students in the last car. As I exit the final cable car, the air is noticeably cooler and thinner, a dramatic change from the early summer heat below. I put on my jacket and walk around. At the top, there's a James Bond museum and a restaurant and then an outdoor viewing area with 360-degree views. There are a handful of people in that viewing area, but beyond us almost all I can see is mountains, a billow of overlapping crags, some shadowed by valleys and highways of snow, sharp angles of barren rock on others, becoming—hundreds of meters below—evergreen-forested slopes.

I walk along the edge of the platform. There are signs detailing which mountains can be seen in each direction. I see Eiger, Mönch, and Jungfrau. The sky is clear in all directions except for one. A low-hanging cloud obscures from view a single mountain in the distance: Mont Blanc.

The nearest I come that summer to Mont Blanc is taking a photo in the airport at a store named after the mountain. I assume when I walk down the terminal gate toward the store that it will be an outdoors shop. It turns out it sells diamonds.

## LOST ON MONT BLANC

Mont Blanc, I read in a tourism guide long before I visit Switzerland, is not only the highest mountain in Europe; it's one of the most visited mountains in the world, with nearly twenty thousand people summiting it each year, math that works out to about fifty-five people per day.

Nearly a hundred climbers die every year in the Blanc Massif. At busy times, local officials perform an average of twelve rescues per weekend, often of hikers and climbers who are disoriented, unprepared, or find themselves injured or in bad weather.

In 2007 two outhouses were helicoptered in and, after being placed at 13,975 feet, began to be emptied, also by helicopter, every few months and sometimes weeks. This was an effort to keep Mont Blanc—once known as the white lady as well as the symbol for modern mountaineering—from becoming, as someone puts it in an article I read, "Mont Noir."

## LOST IN MONT BLANC

> The everlasting universe of things
> Flows through the mind and rolls its rapid waves
> Now dark—now glittering—now, reflecting gloom
> Now lending splendor, where from secret springs
> The source of human thought its tribute brings
> —Percy Bysshe Shelley, "Mt. Blanc," July 23, 1816

## LOST PROMISE

The summer I visit Switzerland, the president of the United States declares that he will be withdrawing the country from the Paris Climate Accord. "As someone who cares deeply about the environment, which I do," he says in his official statement, "I cannot in good conscience support a deal that punishes the United States, which is what it does."

In that same statement, the president explains that the United States will instead increase coal jobs. "We're having a big opening in two weeks. Pennsylvania, Ohio, West Virginia, so many places. A big opening of a brand-new mine. It's unheard of."

## LOST PLACE

Within six weeks of that announcement, an iceberg the size of Delaware breaks off the Larsen C Antarctic Ice Shelf.

## LOST COUNTRY

Four months after that, the United States suffers Hurricane Harvey, Hurricane Irma, and the most significant California wildfires in its history.

## LOST SCIENCE

Five months after that, the Environmental Protection Agency pulls three of its scientists—set to talk on climate change—from the lineup of a Rhode Island conference.

## LOST MAP

In spring 2017, the same spring the Larsen C Ice Shelf splits, the *New York Times* features the three-part series "Antarctic Dispatches," titled "Miles of Ice Collapsing into the Sea," "Looming Floods, Threatened Cities," and "Racing to Find Answers in the Ice."

In the first of those dispatches is an online moving map of Antarctica's ice. Blue lines of ice flow down the Ronne Ice Shelf, the Brunt Ice Shelf, the Amery Ice Shelf, the Shackleton Ice Shelf, the Ross Ice Shelf, the Getz Ice Shelf, and not far from the Cape of Disappointment, down the Larsen Ice Shelf. Even when I set my computer down and stand up to get a drink, I can see, in the distance, nearly every part of the map moving toward the sea.

"The acceleration," the *New York Times* writes, just below its moving map of receding glacial ice, "is making some scientists fear that Antarctica's ice sheet may have entered the early stages of an unstoppable disintegration."

## /LÔST, LÄST/

In 1300 the word "lost" came to mean "wasted," "ruined" or "spent in vain." By 1500 it also took up the meaning "gone astray."

## LOST FATHER

For some time afterward, there was a custody battle over the mummy, Otzi, found in the Alps. Since Otzi was discovered right on the border of Austria and Italy, both countries wanted ownership of his body. Eventually it was decided that Otzi had been located five meters into Italian territory. After that, more court battles raged about who should be compensated for Otzi's find and for how much. Two women separately declared that they had seen the body before the Simons did. One said she had "spit on the body to claim it." Neither woman could verify their claim, though. In the end, Otzi's body was placed in a museum in Italy, in a freezer that mimics glacial ice conditions.

On October 15, 2004—two years before the Simons were awarded nearly $100,000 for their discovery of Otzi—Helmut Simon returned to the same place where he'd found Otzi. It should have been an easy climb—four hours of hiking—but it began to snow once Simon was already on the mountain. Temperatures dropped; unable to find his way, Simon slipped into a three-hundred-foot ravine; his body was found eight days later.

Helmut Simon was not the only one connected to Otzi to have died. Since 1991, seven people who have had close contact with Otzi have died: Rainer Henn, a forensic pathologist, died in a car crash; Kurt Fritz, a mountain guide, was killed in an avalanche; Rainer Hoelzl, a journalist, died of a brain tumor; Dieter Warnick, a rescuer, died of a heart attack hours after Simon's death; Konrad Spindler, an archaeologist, died of ALS; and in 2005 Tom Loy, also an archaeologist, died of a blood disease. The German press has called it "the curse of Otzi."

Simon was once quoted as saying, "Otzi was like our son."

## LOST MANTRA

When you search for what is lost, you need to be careful, it seems, not to become lost yourself.

## /LÔST, LÄST/

In the 1630s the phrase to "lose one's heart" emerged to mean falling in love. In 1744 the phrase "lose heart" began to refer to discouragement.

## LOST STUDENT

One of the friends I travel to Switzerland with tells me that a student of hers once got lost in the Mount Hood wilderness area in Oregon. The student had decided to go hiking on her own one Sunday night in late March. She was supposed to be backpacking with a small group of friends, but in the end, for one reason or another, each of her friends canceled on their plans. Rather than postpone the trip, the student sent an email to a friend mentioning she'd be on Mount Hood and then set off. She'd packed a sleeping bag but left it in her car and set out on foot with only a backpack that carried some clothing, climbing supplies, and a day's worth of food.

When the student didn't come home after some time, the friend alerted the authorities. Her credit card was traced; video footage showed her at a store on the way to Mount Hood, buying shoes and an axe. Search-and-rescue teams were sent out to comb the mountain and surrounding woods.

Six days after she went missing, a National Guard helicopter spotted her. She had run into a whiteout while trying to summit the mountain, then fell forty feet and injured her leg. She'd dragged herself up to an area where she thought someone might see her. Miraculously, they did.

"Were you afraid?" my friend had asked her student.

I am surprised when my friend tells me the student had replied "No," that she knew eventually she would be found.

## LOST VALLEY

If Air India Flight 101 had flown fifteen meters higher, it would have missed the edge of Mont Blanc entirely. After the crash it was determined that a communication error led to the plane hitting rather than missing the mountain. The air traffic controller told the pilot to descend after Mont Blanc. The pilot seemed to think he had already passed the mountain.

Conspiracy theorists, though, wondered if the plane was made to crash intentionally. Former CIA operative Robert Crowley claimed that a bomb had been placed in the cargo area of the plane. Homi J. Bhabha—the father of India's nuclear industry and the chairman of its Atomic Energy Establishment Trombay—was onboard that plane on his way to attend a conference in Vienna. The crash, Crowley and others suggested, was meant to slow India's development of a nuclear bomb.

The plane flying Air India 101 was a Boeing 707-437. It was named Kanchenjunga, after the third highest mountain in the world, an Indian peak that is said to be the home to the valley of immortality.

## LOST IN ICE

Since the Air India 101 crash, the Bossons Glacier has been the site of many significant finds, both fragments of the plane and items it had been carrying.

In 2008 a mountain climber found a set of Indian newspapers there, dated January 1966.

In 2010 a British university student on a class field trip found a blue mail bag that contained seventy-five letters and cards.

In 2012 two climbers discovered another bag of mail, this time a twenty-pound bag of diplomatic mail marked "On Indian Government Service, Diplomatic Mail, Ministry of External Affairs." That bag contained copies of the *Hindu*, the *Statesman*, and Air India calendars among the mail.

In 2014 an almost-fifty-year-old camera was found by another French climber near the site of the crash. That same year a treasure hunter, Daniel Roche, found fifty pieces of jewelry on the glacier. He said the jewelry wasn't as valuable as the unnamed student's find of the hundred jewels the year before. In any case, he planned to keep what he had found.

One year later Roche found an upper thigh and a hand sticking out of the ice.

The same month Roche found those body parts on the Bossons Glacier, three other bodies were found in the Swiss Alps. Two, found on the Diablerets Massif, were a shoemaker and a teacher who had disappeared seventy-five years before. The other body, found on the Lagginhorn, was a German hiker who had died thirty years prior.

## LOST FRIENDS

My friend's student was not the first to be lost in Oregon's Mount Hood Wilderness Area. In August 2010 the bodies of a friend-of-a-friend, Katie Nolan, and her climbing partner, Anthony Vietti, were recovered from the Reid Glacier on Mount Hood, eight months after they went missing, once the warmer weather had melted the snow enough to see them, there, suspended in time.

## LOST CHANCES

A few months after Katie Nolan's body is found, I attend a wedding in Oregon on the edge of the Mount Hood National Forest. I plan to hike Mount Hood myself while I'm out there and even pack, in a carry-on, all my glacial gear: harness, day pack, wool underclothes, headlamp, food, waterproof everything: gloves, jacket, pants, and boots, leaving room only for a bridesmaid's dress and a couple regular changes of clothes.

While we're there, the rest of the wedding party and I stay in a small Scandinavian-themed cabin, the Heritage, that is poised high in the evergreen forest. The cabin is lined in dark wood with blue shutters. On the wall is a framed copy of the Lord's Prayer in Norwegian next to some museum photos from Oslo.

As it turns out, I never make it to Mount Hood. It rains and snows the entire five days that I'm there except for a twenty-minute break in the middle of the ceremony when a slice of light appears through the branches, through the windows of the wooden rotunda where the wedding is held and onto the bride's bare back.

The light doesn't last; as soon as the wedding is finished, it begins to hail—*like rice,* one of the other bridesmaids says, *only harder.*

## LOST ON MOUNT HOOD

Two years after the wedding in Oregon, I do make it to Mount Hood. My friend Kim is visiting the United States from England and offers to meet up in the Northwest; I suggest Mount Hood. On the day of our hike, Kim drives our tiny rental car up a narrow gravel road that looks, at first, like it's a driveway or logging road, at best. But it goes up, curving higher into the woods, and so we follow it.

Finally, we find ourselves at a small parking area. We park next to the one other car there, put on our hiking gear, and start upward, into the woods. We're supposed to be making a loop toward Mount Hood. The shady uphill path moves us that way at first. After thirty or forty minutes, we can see the mountain itself, ahead of us, its snowy outline looming above the trees. I take a photograph of Kim and myself and the mountain. We're wind-whipped and my face is red from the hike, but the sky is clear behind me.

We turn just after that, onto a high ridge that veers back into the denser woods. Perhaps that's where we go wrong. After only a few minutes, instead of continuing to walk up, toward the mountain, we seem to be going down. We walk and walk some more, and I feel certain this must just be a dip before a rise, but there is no rise. When we finally turn around and walk back, we can't find the path up from our map.

Somewhere near the lowest part of our descent, I hear three sharp blasts of a whistle, as if someone is calling for help, but in the dense trees, I do not know where the whistle is coming from. I cannot make my way to it.

## /LÔST, LÄST/

The phrase "losing it" began to be commonly used in the 1990s. When researching the phrase, I come across an English-language-learners' website, where a new English speaker titles their post "Am I losing it?" and asks, "Could you please tell me the best meaning for this sentence? By the way, should I just use it when I'm talking with my friends and family?"

Two people reply to say the phrase means getting angry or losing composure. One woman from England adds that it also means "losing one's marbles" or, sometimes, "losing the plot."

## LOST PLANES

Air India 101 is not the only plane to have crashed into Mont Blanc. In fact, fifteen years before Air India 101 took off—November 3, 1950—another plane crashed into the mountain in almost that exact spot. That plane, a Lockheed L-749a, Air India 245, the *Malabar Princess*, was carrying forty Indian soldiers from Bombay via Istanbul and Geneva to London. There was stormy weather that day. Rather than coasting over the top of Mont Blanc, Air India 245 crashed straight into the Bossons Glacier at 15,344 feet. It took three days before search parties could reach the plane; there were no survivors.

## LOST SAINT

One summer I visit my friend Laura, who has recently gotten married and moved to Europe. She and her husband, Andrew, take me to country pubs and an old fortress; we have a picnic in a meadow. The day I arrive, they pick me up from a small train station in a medium-sized city, the city where Andrew grew up. "Would you like to have a tour?" he asks, and I nod. We wander by shops and restaurants, through a garden, past a river, and then he points out the city's cathedral.

"We should go inside," Laura says, and grabs my arm. We walk right past the paying entrance. "Andrew doesn't believe in paying to visit churches," Laura says, as we push open a side door and walk through.

We wander around the cathedral, past its long wooden pews and cement plaques and tombs for famous people who were buried in that spot. After a few minutes of walking around, and while Laura and Andrew are looking at a tomb and discussing some historical figure, I walk to the very front of the church, to a narrow hallway behind the lectern and communion table and seats for the priests.

There's a wooden door there and it's open. Though I'm not sure what this place is or even if I'm supposed to be there, I step inside.

Inside the room is a small chapel, filled with paintings in woodcut frames, each painting of a different Christian saint. My eye is drawn to a single painting, though. There's greenery around the saint; he has wispy brown hair, wears a cloak, and is carrying a long wooden staff, still bearing the marks where it was sawn off when once a tree branch. A small nameplate below the painting reads "Saint Christopher." A woman standing just behind me looks at the same painting and says to her husband, "He's the saint of lost things, isn't he?"

"No," her husband replies, reading off a small yellow cardstock brochure that he has in his hand, something he likely picked up from the information table in the corner. "He's the saint of travelers," the man finishes, "and also of mountaineers; he once carried the Christ child across a river."

Though I'm not of a tradition that venerates saints, in that chapel, my friends waiting just outside, I kiss my palm and touch the saint's head.

## SAINT OF THE LOST

The real saint of lost things is Saint Anthony. One time, tradition says, he prayed for a lost book to be returned to him and it was. He also had a gift for preaching. One day—years before he'd preach to crowds of thousands—there was no one there to listen and so he went out, by himself, and preached his message to the fish.

## LOST AND FOUND

When I'm in college, a friend invites me to a Pentecostal tent meeting. It's the South, but even so it's an unseasonably warm winter.

Partway through the meeting, the preacher tells the crowd that sometimes God gives signs to lost-people-who-have-been-found by God: sometimes healings, sometimes "physical manifestations."

I don't know what he's talking about, not until the woman next to me leans over and shows me her palm. "Look," she says, pointing at her hand. "That glinting you see; that's diamonds."

I try to look closer, but all I see is water.

# GLACIOLOGY

It's quiet on the glacier—and not the good kind of quiet. It's the long quiet—the quiet that splits you open, leaves you flayed.

I try yelling.

"Matt—"

"Adam—"

"Lydia—"

"Help—"

Nothing. Only low rumblings in the distance and the faint sound of water running, probably rainwater or melted snow, dripping through thin tears in the ice, pushing downward, drips becoming streams, streams becoming wide rivers of glacial runoff, pouring down the base of the mountain, splitting into the glacial valley, cascading into the fjord and then the Atlantic Ocean.

When my eyes adjust to the semidarkness of the crevasse, I size up the hole. I'm hanging midway between parallel walls of raw ice, thick and slanted and buckling in places. There's an overhang two stories above; I see the outline. Shards of snow crack off it every few minutes, fall past me or onto my back and arms. Somewhere above the overhang, there's a shaft of sky. Beyond this, the only thing I can make out clearly is a thin blue line, edging the glacial walls many stories below. Everything else is wet and dim, like the underbelly of a cement culvert in high tide.

It was the summer of the lemmings: the fourth year in the four-year cycle of boom and bust, massive population explosions then sudden and devastating dives; in the course of a few months, sometimes weeks, Norwegian lemmings fall off cliffs; they walk into rivers; they climb onto long sheets of ice and sun themselves to death. In a single area, populations swoon from several thousand to near-extinction.

No one knows exactly why this happens. Some scientists say that the Norwegian lemming deaths are caused by the early thaws, then late spring freezes that melt the tunnels, that put the lemmings on the top of the ice, bare and exposed. Lemmings burrow, create long, low caverns beneath the upper layers of snow. When the snow melts early, they can freeze in a single cold morning. Others say that it's a stress mechanism: there are too many animals in one place, and in their sprint to get away from one another, they run off cliffs, they dive into deep pools of water, they expose themselves to the elements.

In the 1960s, scientist W. B. Quay suggested that in the every-few-year combination of bursting lemming populations and increased temperatures, something inside the lemming's brain becomes unbalanced. Abnormal deposits begin to show up in their blood, and the lemmings begin to move at a frantic pace. These deposits shut off all the normal responses of the animal brain until the lemmings can't eat, can't dig, can't reproduce, can't do anything but move, in frenzied circles, in large groups, until they exhaust themselves to death. "It is mass hysteria," wrote Ivan T. Sanderson in 1944, in the *Saturday Evening Post*, of a summer of the lemmings. "There is no turning back. These timid, retiring animals have lost all their natural sagacity."

I didn't know about the lemmings that day on the glacier, but I'm not sure it would have mattered if I did. All summer my friends and I had been eager to get onto the glacial ice. I'd hiked there once before, but it was the first time for most of my friends, and none of us were from Norway or from places with similar topography. We were driven—it seemed—by the idea of traversing a private plane of ice, of finding ourselves in a landscape that was both swelling and drifting off-center at the same time.

We had gotten up early that day, taken the first bus, and arrived at the glacier before breakfast. It was clear, dry, and sunny and mostly cloudless, that morning of the glacial hike. Unlike the rest of Norway—wet and green—the high glacial mountains are dry, cold, austere: brown grass plains flattened by heavy snows giving way to sharp angular slopes and broad block fields of jagged rocks. When there's no snow or rain or fog, you can see the backside of several summits in the distance, snow crusting over their ridged outlines, then dropping five or six thousand feet to wide rocky valleys. Beyond the tall mountains, miles of high plains and lesser peaks, gray and muted—even in the summer sun—seem to stretch straight to the sea.

Unlike some other sections of the glacier, this one took more work to get to; after we got off the bus, we hiked, for some time, down a single-track trail, over cold knee-deep streams and long washes of mud, before we made it to the perimeter of the glacier, the place where hundreds, maybe thousands, of feet of compressed snow met rock.

At the edge of the ice, our guide, Matt, took a rope out of his backpack and the rest of us began putting on our gear. Once we had our boots, crampons, axes, and backpacks adjusted, he snaked the long diamondback rope across the snowy ground and we all tied in with loose grapevine knots around our waists. Though there weren't any other groups of hikers or climbers on the trail that morning, the snow was dirty and footprinted from hikers and climbers the day or night before. Ours was a popular entrance route to the glacier, only a few hours' hike from the summit where the larger ice forms were.

That's where Matt was taking us, toward those masses of ice. They were blue, even in the distance: mounds, caves, gleaming hallways. Those blue ice forms—the only thing distinguishing the glacier from a snow-covered mountain—would be our first chance to do real, technical climbing: lurching across five-foot wide crevasses, scrambling up and down nearly vertical surfaces, inching over hollow ridges of ice, past deep wells of water, a thousand feet below.

Minus Matt, the rest of us had been hiking together all summer, up and down Norway's western coast. We'd already climbed two of the region's highest peaks, once in a snowstorm. Our last big hike had been Fanaraken, one of the most trying climbs in the area. We could see the glacier from the top; I took a photo of it: six of us in winter jackets in June, standing on the edge of a narrow cliff. Behind us was a vertical drop, thousands of feet down, then miles of white snow, so much snow the background of the photograph looked blank or overexposed.

This day, we were hiking through all that snow: sinking in knee-deep, yanking a leg out, finding some tenuous sense of balance and repeating. The snowfall was recent, had come late to Norway that summer. Snow's always riskier than ice, especially on glacial summits like this one, summits that had an early thaw then a late last storm, hiding the crevasses—long low cracks—that spread like a graph up the mountain.

Sweat and snow had already soaked through my gloves and the cuffs of my pants. It had become rhythmic: our boots sank into deep snow, then we climbed out of it and then sank back in and we climbed back out, stopping between the sinking and climbing only to relash on the metal crampons that were made for ice, not unsteady surfaces. By midmorning no one was talking except for the occasional *this is so hard I want to gouge my eye out* from Lydia, a medical student in the back of the group.

We were finally heading out of the first snowfield—toward all that blue ice—when I noticed it: a brown spot on the ground, just to my left. It was too dark and even-shaped to be mud and too far from the trailhead to be a rock or a patch of dried grass. I stopped and the guy behind me—a lanky twenty-year-old—stopped too. "Are you all right?" he called to me. When I didn't answer, he walked up to where I was standing. I pointed to the spot: "Look at that." He looked for a moment and then crouched down and kicked at the brown spot with the toe of his leather boot, cracking the thin layer of ice that was covering it.

Underneath the snow and ice was a small, frozen animal: a lemming.

In the 1950s, Walt Disney's *White Wilderness* became the first documentary to film the lemmings' dramatic deaths. The climactic scene of that movie taped hundreds of lemmings "migrating toward mass suicide" in Alberta, Canada, brown and white bodies falling, sliding, scrambling over the northern Canadian terrain in a frantic pack. They pushed down and up, and into one another, all tracking a single lemming in front, until they reached a high cliff. The leader jumped: flew gracefully through the air toward the water. Then the rest jumped, and hundreds of lemming bodies, all on tape, were raining into the Arctic Ocean.

Except that in 1958 it wasn't the year of the lemmings. And it wasn't an ocean; it was a river and the filmmakers had paid twenty-five cents per lemming, gathered hundreds of them, put them in cardboard boxes for keeping, then onto a spinning white turntable, which they angled up and down until they had gotten the camera shots, and then they pushed those lemmings to their deaths, off the turntable, toward a long low cliff angling into the water. Some people say that when they watch that movie they see the lemmings hesitate, that they stop slightly, for a second, before they jump.

Norwegian lemmings are the only lemmings whose populations fluctuate randomly, who die in such massive sweeps that they almost don't come back. *They fall from the sky with the rain,* geographer Zeigler of Strasbourg proposed in the 1530s after hearing lemming reports from two Norwegian priests, *and they die when the grass grows in spring.* Collared lemmings, the type of lemmings that live in Canada, in the United States, in most of the places other than Norway, don't live in packs and they hardly ever migrate.

We were gradually making progress up the mountain—following the shallow recessions of each other's footsteps straight through the hard glare of the high alpine light—when all of sudden, without warning, I felt it: a sharp jerk on the rope.

I looked up just in time to see our guide Matt—ten feet in front of me—stumble slightly and then lose his balance altogether, his small frame hurtling straight into the snow. He swung his arm out, grasped for the rope, and threw his yellow backpack behind him, trying to backtrack from his misjudged step. Still, a second later both his legs were gone, plunged in waist-deep.

I stopped, as did everyone else behind me. The snow had been wet all day, but it was the first time any of us had slipped in beyond our knees. "Are you okay?" I called to Matt from behind, holding one hand on the rope and kneeling to feel the ground myself. It was firm, solid, virtuous: an Illinois cornfield in winter, a Wyoming road in the middle of March. Or so I thought at the time.

In one smooth step, Matt leaned forward, yanked his legs out of the snow, and righted himself. He brushed his pants off, then bent down and jammed the metal handle of his pickaxe into the snow a few times, the way he had done every few minutes along the hike—shallow holes, concentric circles—mapping our path up the glacier. He picked up his backpack and started ahead.

"Walk," Matt called back to the rest of us, still standing.

I adjusted my backpack and gloves and then I walked on, veering wide of the hole where Matt's leg had sunk, just past the path of circles he'd made with his axe. At first, the path was fine—steady walking—but as I got to the final circle, the snow suddenly felt wrong, strangely light and loose, less like an icy corridor and more like a frosted-over creek at night.

Realizing what was happening, I threw my body forward, but I was too late. One of my feet started slipping, and then the other, and then I was gone.

Initially scientists believed the Norwegian lemmings' cycles were tied to the cold. In this thinking, the lemmings' migrations were an effort to get away from cold. The theory made sense: the mosses lemmings eat freeze in ice; the thin layer of snow where they dig tunnels and nests cakes over, hard, forcing them to find lower ground. But then, in the 1990s, biologists began to notice an unexpected phenomenon: despite rising temperatures, the lemmings' population peaks were falling. Plows were no longer scraping dead lemmings off the roads, and there were no longer lemmings' carcasses at the edges of rivers, contaminating water supplies. And then in 2001 lemming fossils from the Pleistocene—the Ice Age—were found on an island off the coast of northern Norway near a rocket-launching site.

Biologists began to say it wasn't the cold but the heat that is a problem. In cold weather, the snow's consistency stays relatively constant; the lemmings can camp out below the snowpack, tunneling in deep and eating everything around them. When winter temperatures are too high—spiked by man-made emissions, greenhouse gases, and fossil fuels, the same rising temperatures that crack ice, fracture glaciers, create crevasses—it's then that the snow melts and frozen water floods the lemmings' tunnels until they collapse, drowning some, forcing others to freeze to death, suspended between snow and ice. When the cold returns after a spell of heat, the mosses that the lemmings eat and the lemmings' nests, made with long strips of their own hair, turn to ice. There's no warmth; there's no food. The lemmings are forced to scramble for limited resources, aboveground, over the rivers, and into the valleys. The stress from the overcrowding and hunger causes the lemmings' adrenal glands to swell, creating excessive hormones, increasing inflammation, and lowering lemmings' resistance to disease until the increased hormone production becomes too much and the adrenal glands simply shut down.

"The lemming population is falling and the peaks are disappearing.... A relatively small effect on one particular species is having a broad effect on the system," says Nils Stenseth, lemming researcher from the University of Oslo. The fall of the lemmings' deaths also means a fall of the lemmings' peaks. Ironically, the lack of deaths signals a lack of life.

The ground had collapsed beneath me. I was tumbling then falling through the snow, as if it were a lake on a summer day and I was a bird plunging under the surface of the water. I hit the bottom of the snowpack and went through. The top layer of snow was gone, only sticking to me in wet, heavy clumps.

I was falling, dropping, like ice in a glass, like a stone in a river—clear, narrow, rushing—caught up in the motion or the feeling of motion, motion without action, without intention, just distance and gestures. I slashed my axe through the air but was crashing too quickly to catch the edge, the hole, the faint flash of blue, the walls of ice that were growing then caving in around me, or maybe I was caving in, into the walls, I couldn't tell. Everything was moving too fast: the rope uncoiling behind me, the axe in my hand, the snow all around my head, the glistening walls, a last streak of sky above, damp air, and my body, folding and unfolding like a diver.

A diver with no pool, a faller with no net, no expectation of a net, of a catch, of a fix, of a sign, only air, only atmosphere. Moments layered, space gave way to more space or deeper space. Still, there was no longing there—no, not yet, not in that place—no room for longing, only quiet: the quiet weight of a body falling, of breath inhaled, unable to surface.

Almost immediately following his death in December 1966, rumors began to circulate that Disney had been "frozen alive," his body stored alternately in Disneyland or in Glendale, California, in a large vault, on ice, until some day in the future when scientists could medically resuscitate him. Later some of Disney's biographers suggested that Disney had volunteered to become America's first cryonics patient, meaning his just-dead body would have been injected with antifreeze, cooled to the temperature of liquid nitrogen—negative 321 degrees—and stored in a body bag in a capsule, in "suspension." Disney's family and friends strongly refute this, though they concede that Disney avoided even attending funerals, that he was perpetually anxious about his inevitable demise.

It wasn't just Disney, of course. *White Wilderness* was filmed at the height of the Atomic Age, a mere ten years after Little Boy and Fat Man were dropped on Hiroshima and Nagasaki. For the first time in history, it seemed, it was possible to annihilate the entire human race in just a few swift blows. "I realize the tragic significance of the atomic bomb," President Harry S. Truman wrote in a 1945 statement to the American people. "It is an awful responsibility which has come to us.... We thank God it has come to us, instead of to our enemies and we pray that He may guide us to use it in His ways and for His purposes."

Death had become a feature of the American national psyche; weapons of extermination had become gifts, products of American industry and ingenuity. Photos and video footage of the victims—flash-burned by radiant energy released in heat and light—were classified, hidden. Yet the effects and fallout of the atomic bomb were felt around the globe from Tibet to Greenland and even to Norway's glaciers. In fact, it was with the advent of the atomic bomb that glacial dating changed. It was at this point that scientists began to date glacial accumulation and recession by studying radioactive fallout conferred in two layers from the 1951 and 1963 atomic bomb tests.

"Our films have provided thrilling entertainment of educational quality," insisted Walt Disney of his collection of nature films, including *White Wilderness*, "and have played a major part in the worldwide increase in appreciation and understanding of nature. These films have demonstrated that facts can be as fascinating as fiction, truth as beguiling as myth, and have opened the eyes of young and old to the beauties of the outdoor world and aroused their desire to conserve priceless natural assets."

The myth of the lemmings took off: newspaper comics, commercials, video games. In his 1976 children's book *The Lemming Condition*, Alan Arkin tells the story of a young lemming named Bubber who is the only lemming to question the rest of the lemmings' plans to jump into the sea. "What do you think it's going to be like?" Bubber asks an adult lemming, Arnold. "How should I know?" Arnold responds... "radiating peace, calm and disdain."

In a 2008 bid for the U.S. Senate, Oklahoma candidate Andrew Rice took the myth even further, using the lemmings scene from *White Wilderness* as part of his advertising campaign. "Washington politicians," the narrator of the ad starts as the brown lemmings begin moving through the brown grass and over the rocks, "are a lot like lemmings. They follow their party, even if it's over a cliff." As the lemmings start falling into the water, the narrator begins talking about Rice's opponent, Jim Inhofe, and all his political stands "supporting big business, shipping jobs overseas, against body armor for our troops." "It's time to change direction," the narrator finishes as the video shows dozens of dead lemmings floating in the water.

Whatever Disney's reasons—if he had them—for perpetuating it, the myth of the lemmings became big: bigger than *White Wilderness* and bigger than a ploy against collective bargaining or an admonishment about the perils of groupthink. The story of the lemmings, it seems, came to offer a measure of control, a reverse moral where rugged American individualism alone plays the trump card against the terror of the long edge.


Five minutes. I feel nothing, nothing except my right leg pounding like it has a heartbeat. The tip of my axe is dark and wet, and there's some blood gathering around my sock; my pant leg and my shirt are torn from the axe; one metal crampon hangs off my ankle; the other is split in two. Otherwise I'm fine, or numb; I'm not sure. I'm sweating and shaking, and I don't want to take off my gloves to see whether the tips of my fingers have begun to turn blue.

I heard it first, before I felt it—the weight of my backpack and my body and the fall—snapping hard against the single knot that held me. I lurched, the rope burning against my legs and my waist as my body moved forward and the rope yanked backward, careening between one wall of ice and another. But then, as suddenly as it began, it stopped: I stopped, faceup, hanging in midair. And then it was quiet.

It had been quiet ever since—and more than anything, more than the height, more than the thin rope or the overhang, it was the quiet that worried me. Needing to do something—anything—I decide to swing the rope I'm hanging from back and forth in hopes of gaining momentum and wedging myself against one of the walls.

I hold the rope close to my chest with one hand and begin unleashing things from the side of my backpack with the other: some food, a couple of bottles of water. One at a time, I shed things into the dark below me. When there isn't anything left to drop, I rehearse the plan in my head and pick a spot on the wall that I'll aim for. I swing back and forth three times, before I decide to go for it. I grab the rope harder, bend my knees to my chest, and then hurtle toward the wall in front of me, slashing my axe above my head.

I'm too far from the wall to get any traction, to make meaningful contact; I hit the ice, hard, but I bounce off it as a slab of wall breaks and barely misses my back and falls into the darkness below.

I watch it drop, then I put one hand just under the knot and stretch my other arm straight out, to keep my balance. I don't move, I don't yell. I just hang there shivering, wondering, if I wait long enough, whether the high canyon walls will come together like stairs, will lift me.

I should have been prepared; we prime ourselves for accidents out there, practice clotting wounds, treating frostbite, splinting bones with long, narrow hemlock branches. In my wilderness guide training, we rushed into cold rivers and let ourselves go, into the thrashing current, so someone else could throw out a rope and feel what it's like to save somebody, or to not try hard enough and to watch a person begin to drop, down, into the shimmering deep.

I should have been prepared, but I was not. I hadn't banked on seeing the lemming. *How could I?* I would ask myself later, more as an excuse than a question. When I saw the lemming, that day on the ice, it didn't look stuffed, like I would expect from the dead or the dying, or breakable, like a paperweight or a pool ball. Shorter than a field mouse but thicker, it was light and wet, like an infant still in fluid, and trembling—though that may have been the wind, or me, pressing up against it, shivering. Its eyes were dark, but its ash-brown hair was caked in a thin layer of snow.

I crouched down next to the lemming, as soon as I realized it was an animal—not some dirt or a rock or piece of trash or something else someone had left behind on accident, dropped from a pocket, out of a backpack, never recovered—and touched it lightly with my glove. It shook slightly. I didn't know what to do: to cover it back in snow, to warm it, to wrap it in a hat or jacket or my bare hands, to see if it might be revived.

I did none of these things. "Walk," Matt yelled back to the rest of us. The whole line was stopped by this point, waiting for me as I looked at the lemming. Behind me, my friend Adam hiked a few paces back to his place in the rope-line and pulled the rope tight between us. I stood up too, dusted the snow off my pants and gloves, and then I moved on, leaving the lemming lying in the snow, exposed.

As I began walking up the glacier, though, I saw another brown spot to my left—dark, even-shaped, just below the ice—and then there was another brown spot just ahead of me. Soon there were half a dozen brown spots lining the path, breaking through the surface. We were surrounded by dead lemmings, shaggy, gaunt, and encased in ice.

# Glaciology

In 2006—partway through my first summer in Norway—a team of scientists, under the leadership of Ohio State glaciologist Lonnie Thompson, set out to analyze one of the world's most remote glaciers, the Himalayas' Naimona'nyi glacier, by collecting ice samples from the infamous 1951 and 1963 radioactive fallout layers.

Thompson—who has traveled all over the world studying high-altitude glaciers and is renowned for having spent more time than any other living soul at over 20,000 feet—expected some change in the glacial ice. Patagonia glaciers, after all, had retreated almost one kilometer in the prior two decades, Swiss glaciers had lost 500 meters in the prior three years, and five different Norwegian glaciers had receded over 100 meters each in just the prior year. At 25,000 feet—looming above the world's highest plateau—Thompson and his team assumed rightly that the ice would likely not be the same as when it had last been measured.

What they didn't expect is that the atomic fallout layers would be completely gone, that there—in the center of the universe, the place that Hindus call the *devatma* or god-souled land—all that ice would have already melted.

Sometime later a friend would ask me why, after that day on the glacier, I began to care so much about the lemmings. What he meant to ask, I think, is why—those days in Norway and then even once I had moved home—I got obsessed with lemmings, why I taped color photographs of lemmings on the wall and why I talked about lemmings when we were scraping paint from the back of the house and when we were lying, stretched out slim, in separate seats on the city bus. I checked out a small stack of rodent books, underlined important phrases in life-cycle field reports, but I never did answer his question. By the time I thought of something to say, we were no longer standing outside in the same cold dawn.

You see, I had once imagined myself constructing an alternate life in Norway: painting houses, buying groceries at the local shop, walking along the green coastline, watching the early morning light sweep up the cool wet night. It was a measured life, a controlled life, and not a real one. There is no night in the summertime in Norway, though sometimes the afternoon light, all at once, is blinding.

That day on the glacier, I heard Adam first, then Lydia, both muffled. "You're okay."
"We've got you."
"Get me out of here," I yelled up.
"We're working on it," Adam called back, again muffled.
A few moments later there was a sharp pull on my waist, and then snow was falling around my shoulders, chipping off the open side of the crevasse, and sliding down in sheets as my friends Adam and Steve yanked the ice-caked rope over the edge of the hole. Suddenly the rope and my backpack and my body began to lift, one arm's length at a time, up toward the light and the others' voices and the long plane of ice and snow above.
And almost just as suddenly Steve was yelling, "Your arms—give me your arms," and I could see a widening patch of sky and someone's blue fleece jacket, bright and matted, over the last lip of the crevasse. I was reaching, scrambling, pitching myself toward the surface, and someone was grabbing my arms and my backpack and yanking me up, away from the hole. Then I was being dragged through the snow and out into the open air, more air than I could take in, and I was lying on my back, on the sun-spotted ice, heaving.

# ABOUT THE COLLECTION

All our lives are collections, curated through memory.
—RICHARD FORTEY

## EXHIBIT 1

Norway's Breheimsenteret—like the two other glacier museums perched on the edges of the Jostedalsbreen, the largest glacier in continental Europe—was an architectural feat of sorts. Intended to resemble two ice towers with a crevasse between them, the thousand-foot, $15 million building's logic—like much of Norwegian architecture—was imitation, was representation: stone walls, wooden doors, large-paned reflective windows that, on a clear day, mirrored the mountains back at themselves. The shake-shingled roof of the museum—split in two and merged by a single center-stripe of paneled glass—arched upward like a bell curve, a narrative arc, reaching for a seemingly endless expanse of snow and ice in the distance.

Built in 1993, the Breheimsenteret was conceived as part welcome center, part museum: a crosswalk, if you will, between the local arm of the glacier and the whole Jostedal valley just to its east. There were ice displays, geographical relief maps, and large, nearly wall-sized photographs of the glacier. There was an information desk where college-aged museum staff in matching collared shirts fielded questions about ice flow and regional flora and fauna. There were guides on hand ready to book kayaking tours on mountain lakes, rafting trips on nearby rivers, or walks and climbs leaving straight from the museum, veering three kilometers northwest around the Nigardsbrevatnet Lake—a lake that less than one hundred years ago was part of, and not yet the edge of, the glacier—and straight onto the glistening, aqua-tinted ice. There was a restaurant too and a souvenir shop, beside films, images, and signs about endangered animals and plants, glaciation cycles, and the long and little ice ages.

Though the museum, like the glacier, was inconvenient to get to—a thirty-kilometer drive from the nearest village, several hours from the nearest major city, on a one-lane two-way road that was impassable for much of winter—in the summer months of late June through early August, busloads of travelers made their way from Bergen, Oslo, and cruise ports all along the coast, through evergreen-lined valleys, past dozens of rural villages, hundreds of high mountain farms, and down the country's only four-kilometer toll road to the not-quite twenty-year-old glacier-shaped building that still, in those days, was as much a testament to the marvel of modern tourism as it was to the cultural history of the Jostedalsbreen region or the shifting science of a never-ending, ever-shrinking glacial landscape.

### EXHIBIT 2

"In contrast to the souvenir," writes Susan Stewart, "the collection offers example rather than sample, metaphor rather than metonymy. The collection does not displace attention to the past; rather, the past is at the service of the collection.... The collection seeks a form of self-enclosure which is possible because of its ahistoricism. The collection replaces history with classification, with order beyond the realm of temporality. In the collection, time is not something to be restored to an origin; rather, all time is made simultaneous or synchronous within the collection's world."

## EXHIBIT 3

Museologist Jeffrey Abt suggests that the first contemporary museum, in the sense of being a place for the "systematic collection and study of evidence," was Aristotle's. Of course, the word "museum" predates even Aristotle; the Latin *musea* comes from the Greek *mouseion*, temple to the muses, or a space intended for the study of the arts. Both Plato's library in Athens and Ptolemy's library in Alexandria were called *mouseion*, although Plato's library—like most Greek *mouseion*—was more interested in philosophical conversation than in collection, and despite its inclusion of rooms for study of various disciplines, by most historians' accounts, Ptolemy's space seems to have been more of a glorified research institution than a modern museum.

Aristotle's museum was different. In the fourth century BCE, Aristotle had traveled with his student, Theophrastus, to the Aegean island of Lesbos, home of the poet Sappho and, according to myth, the head of the musician god Orpheus. On their island trip, Aristotle and Theophrastus not only studied but also collected plants and historical artifacts, and they brought these back to Aristotle's *mouseion* in Lyceum. The Lesbos study was not the end either. Aristotle built up a community of students and scholars around him and, amongst this community, promoted his collection of certain historical, cultural, and natural artifacts. Theophrastus thought Aristotle's museum collection significant enough to list instructions for its completion in his own will. *Build a bust of Aristotle*, he instructed the will's recipients. *Replace the tablets that contain explorers' maps* and *make a life-sized statue of Nicomachus.*

Though the term "museum" gradually fell out of vogue, for the next several centuries after Aristotle and Theophrastus wealthy individuals and families and then private offices of the church and government—from Charlemagne to the Medici—continued to build significant collections of art, artifacts, and other cultural and historical objects.

## EXHIBIT 4

"The thrill that students or children express on first seeing the exhibits in the galleries at a great museum never fails to impress me," writes D. V. Proctor. "It is a moment when eyes, hearts and minds, if not ears, are alert and receptive; when energies are stimulated and rearing to be unleashed. Even the most hardened secondary school boy or girl admits to interest being aroused at this moment, even if this is a mere flash in the gloom."

## EXHIBIT 5

It was late summer the morning that we left for the Breheimsenteret and the glacier, but it was cold enough to see our breath rise like smoke in the low mountain air. We were waiting for the bus—my friends and housemates, Luke, Sam, Tom, and I—watching for it on the single blacktop road that led from the sprawling guesthouse where we were living due west to the fjords and the Atlantic Ocean and due east inland toward the whitewashed glacial tips of the high Norwegian mountains.

The guys—unshaven, except Tom—were dressed warmly, in wool socks, waterproofs, boots, and fleece sweaters. My own light jacket rippled in the ocean wind. Luke and Sam sat on the edge of the road, backpacks and water bottles cast behind them onto the freshly paved walkway. Tom—blond and boyish—stood, eating a sandwich. A small pack of birds scattered as I stepped under the concrete bus stop overhang to get warm, their gray feathers sweeping off the ground, bearing against the dew-soaked air, and then taking flight.

It was Sam's first time on the glacier; a British economics student, he was going for the views and for the climb. He'd mentioned it more than once in the weeks before, while we'd walked up and down other mountains and hikes along Norway's western coastline, eaten meals together out on the grassy lawn, and fallen into a friendship that was fast and solid.

Luke, Tom, and I—all in our twenties and all friends for a few years—had been to this section of the glacier three or four times before, but we'd never once ventured into the museum on its edge, never studied its exhibits, spent the day in its collections, or asked what it might have to tell us.

## EXHIBIT 6

"Breheimsenteret" translates in English to "center for the home of the crevasse." It took four years after the Breheimsenteret was built until it was authorized as an official visitors' center. Twelve years after that, in 2009, the land surrounding the center and bordering the Jostedalsbreen was declared Norway's first national park for biodiversity, the Breheimen.

## EXHIBIT 7

In 2011 a man's coat was found at the bottom of a melted glacier in the Breheimen. The coat was made of wool and woven with a diamond twill. Scientists dated it to 300 CE, the oldest coat in Scandinavia. "Without close attention," the Archaeological Institute of America wrote, "many of the artifacts that emerge from melting ice will be lost—decomposed or washed away—before they can be studied."

## EXHIBIT 8

Beginning in the sixteenth century, a general interest in collecting and collections was revived. Perhaps in a nod to the *mouseion* of classical antiquity, Renaissance Europeans began building studiolos, gallerias, and "cabinets of curiosity" right inside their homes—often adjacent to their bedrooms—in order to hold varied collections of relics, plants, cultural and archaeological finds, and other objects of interest. These cabinets quickly gained ground in the popular imagination as representing, at their best, a microcosm of the whole world but one in which the cabinetmaker could and would propose his own "natural" order.

"The cabinet of curiosities, or the Wunderkammer," say curators of one such collection, "was designed to facilitate an encyclopedic enterprise, the aim of which was the collection and preservation of the whole of knowledge." This knowledge first followed classical thinkers like Aristotle and Pliny the Elder but gradually began to mark its own paths. "Over time," the curators add, "these activities

About the Collection

began to reveal new truths in conflict with the tenets of classical doctrine. As a result, they began to undermine the established authority of the ancients, thereby paving the way for new methods of 'scientific' investigation."

The most famous of the cabinets of curiosity was created and curated by a Scandinavian: the seventeenth-century Copenhagen physician Ole Worm. Royals and dignitaries came from all over Europe to see Worm's "Collection of Curiosities" that included runic texts, taxidermied birds, human skulls, deformed fetuses, tortoise shells, dried plants, sculptures, and a huge assortment of other items housed in a couch-lined "wonder room" that Worm had built. Like many of his contemporaries, Worm not only collected but also cataloged his works into a book. Divided into the kingdoms of nature (mineral, animal, and plant), Worm's three-edition catalog included engravings of and notes about the collection and offered his personal interpretations on certain phenomena; *lemmings do not fall from the sky*, for instance, Worm noted about the popular Scandinavian myth concerning the origin of these rodents, *they're blown by the wind onto the land*. This book was titled *Museum Wormianum*.

## EXHIBIT 9

"I have collected various things on my journeys abroad," wrote Ole Worm, "and from India and other very remote places I have been brought various things: samples of soil, rocks, metals, plants, fish, birds and land-animals, that I conserve well with the goal of... being able to present my audience with the things themselves to touch with their own hands and to see with their own eyes."

## EXHIBIT 10

When I was twenty-two I cleared my schedule to visit the Chicago Art Institute every Tuesday, the day it was free for students. I had decided to study the museum week by week, to work my way through it one collection at a time.

The year before, my older brother took some of my friends and me to the institute one cold Saturday afternoon in the late fall when we were visiting him from our small university several hours north. It snowed on the car ride down, and since none of us had brought winter jackets, we borrowed old clothes from my brother and his roommates. I wore a red wool sweater that he had received for Christmas, then accidentally washed on hot. The body of the sweater was short, but the sleeves were several inches too long.

We walked that day through exhibits of Japanese textiles, medieval armor, stone and metal sculptures, and busts of Seneca, the Archangel Michael, and a lion attacking its prey. We stopped for a long time in the modern art gallery before Georges Seurat's pointillist painting *A Sunday Afternoon on the Island of La Grande Jatte—1884*. My brother and I had our photograph taken, backs to the painting, in the same position that Matthew Broderick stood in for the film *Ferris Bueller's Day Off*, our profiles obscuring several hundred of the dots, blending in with thousands of others.

When I went to the museum on my own, a year later, I rarely made it past the two bronze lions that edge the museum's front steps. I studied the lions in great detail, making photographs and sketches from every imaginable angle, sometimes facing the arched doors and windows of the building, sometimes looking out at Michigan Avenue and the rest of the city. I thought about the lions often, about their sculptor Edward Kemeys, and about the names that he had given them, *In an Attitude of Defiance* and *On the Prowl*. I found strangely consoling the way the tails of the lions—pointing to the museum—were bronzed-colored, though the bodies had both turned a perfect shade of oceanic green.

## EXHIBIT 11

On October 8, 1871, a hot dry day, the central business district in Chicago caught on fire. Much of the city—constructed of wood—went up in flames within a few hours. The original Chicago institute—then a four-year-old studio and art gallery, the Chicago Academy of Design—burned to the ground.

For several years after the fire, the academy almost didn't make it. Eventually, however, the academy consolidated with another failing school and, thanks to the timely Chicago placement of the World's Fair ten years later, opened its current building, changing its name to the Art Institute of Chicago.

## EXHIBIT 12

Norway has not just one but three glacier museums, all of which were built in short succession. The Bremuseum opened in 1991, the same year the Jostedalsbreen National Park was established. In the next two years the Jostedalsbreen National Park Center and then the Breheimsenteret followed.

The Bremuseum was designed by celebrated Norwegian architect Sverre Fehn to resemble a rock left behind by the glacier's retreat. The walls are cast-concrete; there are skylights all along the top ridge of the roof, and the tunnel-like entrance is flanked by two sets of stairs, aligning exactly with the mountains as you see them from the open-air second-story rotunda. There are twenty-four exhibits inside, including several elaborate models of glaciology in 1 to 30 and 1 to 50,000 scales. There are charts and photographs showing the history of the region from its formation and recurrence in various ice ages to the wedding parties and bootleggers who made its first recorded crossings. There's even an exhibit meant to resemble a glacier meltwater channel: the mockup of the underside of a glacier. It's painted in brilliant blues and drips real water on visitors. Grinding noises, mimicking the sounds of moving ice and stone on the bottom of the glacier, blare through a speaker inside the tunnel. Two mannequins dressed in climbing gear wait near the entrance.

The Jostedalsbreen National Park Center is what its name suggests: exhibits and information all about the park and plants and animals, maps of the glacier, and photographs of wildlife. There is a geological commons attached to the outside of the center—hundreds of different kinds of rocks, sedimentary, igneous, and metamorphic, all poised on two- or four-legged metal stands—and a large botanical garden with five hundred species of wildflowers. Inside, the building is half stave hall, half stone auditorium, symbolizing, as one Norwegian writer puts it, "the encounter between the past and the future." There are carefully labeled rocks lining glass cases along the walls and a large exhibit of taxidermied animals, including a full-sized moose that appears to be running directly from a wall of windows, toward the rest of the exhibits.

## EXHIBIT 13

By that morning my friends and I were waiting for the bus for the Breheimsenteret, I'd already been to both of the other Norwegian glacier museums. A friend and I had stumbled upon the park center the summer before, after planning to get to one side of the glacier but accidentally taking the wrong boat for five hours toward the other side of the glacier, before realizing our blunder.

"I can take you to a car ferry," the boat captain said as he walked up to us, map in hand, after hearing talk on board that some American girls had ended up on the wrong boat. "The ferry will take you to the north side of the glacier, rather than the south; I think, from there, you'll be able to get a bus." I agreed and thanked him, and we soon joined two other passengers on a car ferry void of cars. We got to the park center after hours that night and so camped a mile down the road, beside a small stream. In the morning our shoelaces and water bottles had frozen solid.

The day I visited the Bremuseum, a guide there told me that the museum's slogan is "a place for curious people" and suggested three things I might find interesting: a three-hundred-degree panoramic film of the glacier and the local region, a several-hundred-year-old melting piece of glacial ice set in granite near the front of the museum, and a display about Otzi, the world's oldest mummy, discovered on a glacier in the Alps in 1991, after the ice that had held him for hundreds of years disappeared.

I took the guide's advice: watched the movie, touched the shrinking piece of ice, and then I stood before Otzi for a long time, comparing the plastic skeleton on one side of the exhibit—faded yellow bones, open mouth, a sunken spot in the skull for eyes—to the mannequin man on the other, dressed in furs and leather, sandy brown hair lightly touching the back of his neck.

I brought along a notebook and scrawled pages of notes. I bought a book on Norway's glaciers, a brochure on the national park, a map—outlining, in red, the various routes to get to the glacier—and, for twenty-five kroner, a blue photocopied guide to the museum's exhibits.

4.6 billion years ago [the first page of the guide begins], the Earth was part of an inferno of colliding asteroids…

40 million years ago, the Earth was warm and the atmosphere contained a high percentage of greenhouse gases…

20,000 years ago, at the end of the Last Ice Age, large parts of the northern hemisphere were covered by enormous glaciers…

We now travel to the year 2040. What has become of the world?

## EXHIBIT 14

In 1997 Ada Louise Huxtable wrote, in the *Wall Street Journal*, of the Guggenheim Bilbao Museum, "The container and the contained, the art and the architecture are one thing, made for each other; nowhere else do all of the arts support and play off one another in a unified aesthetic that so fully expressed the twentieth century." In another article, a year later, about the Scandinavian museums including the glacier museums, Huxtable described the ideal museum as "full of the promise of aesthetic and poetic power."

## EXHIBIT 15

Of the Bremuseum, famed architect Kjetil Troedal Thorsen said, in 2010, it "communicates like a poem with its surroundings."

## EXHIBIT 16

Though no one actually knows how the Chicago Fire of 1871 began, city lore argues that the fire started when Patrick and Catherine O'Leary's cow kicked over a lantern in their barn, setting the neighborhood and then the city on fire.

## EXHIBIT 17

The first public museum, the Ashmolean, opened at the University of Oxford in 1683. The museum's origins are traced to a partnership between renowned English botanist John Tradescant the Younger and politician Elias Ashmole. Long interested in collections, Ashmole visited the Tradescants' sprawling Wunderkammer of books, coins, costumes, weaponry, animals, and plants (alternately called the Ark and the Museum Tradescantianum). Ashmole was so impressed with the Tradescants' collection that he offered to pay for and publish a detailed catalog of the museum's contents.

Tradescant the Younger accepted and began a somewhat shaky partnership with Ashmole that ended with his 1662 will deeding his entire collection and property not to his wife but rather to his *good friend, Elias Ashmole*. Fifteen years later—after surviving an extended trial about whether the inheritance was "swindled" or rightly won—Ashmole added his own prints, coins, and metals to the Tradescants' collection and offered up the newly deemed Ashmolean Museum to his alma mater, the University of Oxford, on the conditions that the collection would be housed in its own building and that the building would be open for public viewing. After six years of construction, the museum opened on Oxford's Broad Street on May 24, 1683.

Benefiting from the eighteenth century's quest to advance knowledge through reason and science, the public museum model quickly took hold, and a slew of museums were built—or newly opened for public viewing—around Europe and the world. In 1734 the Vatican's Capitoline Museum opened; in 1759 the British Museum was built; in 1765 the Uffizi opened to the public and, in 1793, the Louvre.

Some people say a crisis of authority prompted the rise of the public museum; museums both accepted and civilized the lower classes, placing visitors in the role of the viewers and requiring them, as one scholar notes, to take on "the ideal spectator pose." Museums, it seemed, offered an education about socially acceptable civic behavior by both informing and constructing a public at the same time; museums secured a belief in the centrality of evidence, interpretation, and classification. *Everything is collectible*, the museum's presence seemed to suggest, *everything is within our grasp*.

## EXHIBIT 18

Historical mythology often recounts the burnings of the *mouseion*, and particularly the Library of Alexandria. Though scholars don't know if there was just one fire—said to be started by Julius Caesar in 48 BCE—or several smaller fires during the four-hundred-year period between Caesar and a decree against the institution by the pope in 391 CE, the museum's destruction became a symbol for the shutting down of scholarship and the loss of cultural knowledge.

## EXHIBIT 19

When I first began teaching, I told my students that essays were a lot like museums, that creating an essay was about gathering artifacts, about doing careful and thoughtful research on those artifacts, and about organizing them into exhibits and collections. *Selection, preservation, presentation*, I'd tell them, *the art of finding something to say and saying it rightly*. My own professor in graduate school, Beth, had asked us to study Los Angeles's Museum of Jurassic Technology as an essay, a collection of found materials; there was a tie, she suggested, between the form of that museum and the nonfiction we were writing.

Like my professor, I had my first classes of students look through the museum's online displays of Russian tea rooms and dogs of the Soviet space mission, but then I sent them to traditional museums—history, science, art, and archaeology—and had them take notes, draw elaborate maps, and analyze structuring principles. As a class we visited halls of statues, photography galleries, a planetarium, a cemetery—*a collection of stars, a collection of bodies*, I told them—and mimicked their designs in our writing.

I wanted my students to see the essay as not only art but also science. I wanted them to have something to hold onto—a consolation, a reprieve, an architecture—when the onslaught of moments, ideas, and words left them circling, breathless.

## EXHIBIT 20

The drive to the glacier involved two bus transfers and an hour-and-a-half wait in a town slightly larger than the village where we lived. When Sam, Luke, Tom, and I finally boarded the last bus, I asked the guys where they wanted to sit and they automatically pointed to the back: behind all the hikers, climbers, couple of students we knew, many that we didn't, behind the luggage racks and the extra door and the reclined seats of the people ahead. There were already several Norwegians onboard, along with a group of twenty or so French athletes in their late teens or early twenties, wearing matching red ski suits.

It was early and so most of the bus was quiet. Luke and Tom, across the aisle, slept. Sam and I talked for a while, but then we too succumbed to the silence, watching the landscape outside the bus windows change. Even after a month in Norway for Sam and several stays in Norway for the rest of us, the views outside the bus windows were startling: the bright-blue fjord stretching like a channel between evergreen heights, precipices of rocks, whitewater and fishermen in the snow-fed rivers, mountainside meadows, and the glacier beyond, fading into the cirrus sky.

The bus traveled in and out of the countryside, all along the one highway that could take us to the glacier: Norway's winding, two-lane 604. The bus passed an evergreen forest, high hills, crossed and crossed again a single river in a straight-north course from the fjord to the glacier. The bus hurtled through tiny villages—blurs of houses and farms and sometimes a single store or gas station—past ships loaded with orange and blue metal-sided shipping containers, past a camping park and a small sign advertising, in English, rafting and zipline tours.

Finally, past a steeply sloped valley dotted with summer houses and sheep farms—hiking and climbing gear clanking on the floor, water bottles, cameras, and ropes on people's laps, crampons and axes strapped to backpacks—the bus turned down the winding four-kilometer toll road that leads to the Breheimsenteret.

It all looked the same; so much of it looked the same. Perhaps this is why I didn't notice, why I didn't see.

## EXHIBIT 21

"You see what you want to see," my father tells me one day. We have just flown across the country for a family vacation, and I look down to discover the driver's license that I have been carrying around for months has misspelled my name and I haven't once noticed.

My architecture professor in the only drawing course I will take in college will tell me the same thing, except he will say, "You choose what to see." He is trying to teach the eleven students in our class—all sitting on high stools with white drafting tables and blank sheets of paper—to draw models and buildings in the same way, by not seeing the object itself but the blank space surrounding it.

I do not perfect drawing the blank space. In that room, I never want to draw around the models or pieces of fruit. I see only the high narrow skylights and the open door behind me, pitching sunspots, warm winter light, onto my pencil box.

## EXHIBIT 22

My first year in Norway, I bought a book whose jacket copy compared a collection of lyric essays to the Museum of Jurassic Technology: this book "is to essays what the Museum of Jurassic Technology is to gallery dioramas."

A few years later I read an endorsement by Lawrence Weschler—author of *Mr. Wilson's Cabinet of Wonder*, the story of the Museum of Jurassic Technology—on another book jacket for a different essayist. "Downright near infinite," he writes, "at any rate, the good fortune of a city blessed with such antic chroniclers."

"To build a museum," Sverre Fehn once said in an interview, "where there is no object, the museum becomes the object and architecture the story. It is a search into the surface of the earth.... The invisible becomes visible."

## EXHIBIT 23

In 1973 Swedish museologist Ulla Keding Olofsson declared that the increased demand for museums "has been so rapid and has reached such a level that museums have to now turn down requests for service." She continued, "So far as it can be foreseen, the factors responsible for increased demands on museums are likely to continue to prevail indefinitely."

## EXHIBIT 24

By 1990 curators and researchers had begun to complain that many museums, but particularly the natural history museum, focused on education and display over research. Their habitat dioramas and "nature-faking," as one curator put it, are more *products of imagination than science*. Museums around the world began to change their names, from natural history museums to museums of science. Patrons, these curators argued, wanted experience, not collection; they didn't have the patience for reconstruction, for that certain sort of seeing.

## EXHIBIT 25

In a 2012 museum symposium, "The Elephant in the Room," Oxford curator Darren Mann declared that for the cost of one painting the Ashmolean had recently purchased (£7.8 million), the United Kingdom's entire collection of entomological specimens held outside museums and universities could be rehoused.

## EXHIBIT 26

In 2013 Oxford's Museum of Natural History held another symposium, "Crap in the Attic?", which tried to respond to the threat of the loss of natural history museums altogether.

## EXHIBIT 27

The bus jerks to a halt and there's a blur of movement. The red ski suits are shuffling and grabbing long black gear bags from overhead, hoisting them on their shoulders or under arms and filing off the bus. The Norwegian students and day tourists are gathering their things too, reaching under seats, zipping backpacks, capping bottles, and tying bootlaces. There's a glint of sun through the window. The bus driver leans into the aisle and stretches his legs.

I begin getting up too. I collect my notebook, pen, water, my camera, and my extra clothes, and I stand up and walk a few paces down the narrow aisle of mostly empty seats toward the low tree line of just-green mountains in the distance, toward the long deep lake framed by the silt-covered beach, toward the ashen-white glacier, thick and vast and filling the valley like a renegade river, like a flooded dam, toward the hikers and brightly colored climbers that I know will already be out, ascending the mountain in long slow lines, toward the day and the museum.

I get to the bus door and realize the guys haven't followed. As I turn around, Sam calls from the back, one hand resting on the edge of my seat, the other gesturing out the now-clear window. "Is this it?"

I look, but I cannot see.

I leave my bag on the front seat of the bus and step out into the warming air; it smells of smoke and pine, and even though it's mostly overcast, the light reflecting off the snow in the distance stings my eyes.

I move through the crowd, past the skiers and the hikers and the families snapping photos and milling about the blacktop parking lot. I walk past the buses and parked cars and a thicket of dark pines and spruce, some scraggly, some full and bursting with branches, all shadowed by the mountain, all rising from the sidelong horizon, from the empty valley in front of the bus.

There are no stone walls, no long wooden door, no large reflective windows, mirroring the mountains back at themselves. There is no shake-shingled roof, split in two and merged by a single center stripe of paneled glass. There are no exhibits or information desk, no photographs of the glacier, and no discussions of its recession. There are no tour guides in matching black polo shirts, no signs, no ice flow or flora and fauna charts labeled in English and Norwegian, German, and French, no lines of people waiting to buy ice cream or an extra pair of gloves or a map of the mountains.

Instead, just past the spot where I stand, there are blackened hunks of wood and concrete and piles of dirt and rock. There's a level spot of ground where a building's foundation once was, and in the air, cut off at the fourth rung, a metal casing and some remnants of glass.

When the wind picks up, all around the parking lot move loose pieces of ash, like a thin coating of snow, like scraps of paper.

## EXHIBIT 28

The year before he died, a friend and former student of mine told me about Los Angeles's Getty Museum. It was the spring before I first went to Norway. We were standing on a brick-lined sidewalk outside a small library; flowering dogwoods and hydrangeas edged the building, spilled over onto the path. He told me, that day, that the white walls of the Getty gleamed in the sun like the snow on a mountaintop before it had been walked on, on a new morning, in the winter air, in the early light.

I watched him intently, nodded along, but I did not realize at the time that this was a metaphor. So he told me the story again, except that this time there was white pavement, white flowers, white statues in reclining positions. *It was white as clouds on strings,* my friend told me, *in the valley of angels.*

## EXHIBIT 29

One year winter wildfires begin to spark across Norway. I watch a Norwegian news broadcast of these fires, gray plumes of smoke drifting over the water, across rocky plains. The footage is taken from a helicopter. Occasionally you see the rounded outline of the helicopter's window on the screen's edge, and then, all at once—following the smoke's same horizontal line—bursts of red. "Freakish winter wildfires in Norway," wrote the United Nations Office for Disaster Risk Reduction, "have underlined the importance of adopting a multi-hazard approach to disaster risk management in an era of changing climate."

## EXHIBIT 30

A neighbor saw it first—flames—shooting up the side of the building, using the wood roof as kindling, torching the exhibits. Townspeople came out and firemen; the museum's curators were called, but there was nothing they could do so they stood back, a hundred meters or more, leaning on cars, crouched along the wet pavement of the highway, and they watched smoke drift over the high glacial mountains as the fire blazed on, like a burning bush, like the apostles' tongues, like a photon emitted from the sun.

## EXHIBIT 31

"You feel like you are seeing everything now. Nothing was happening, and now everything is happening. Why does your sight seem now so sharp and clear?"—Craig Arnold

# PAULING'S CORE

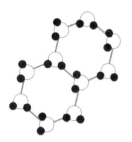

In 1935, when he was thirty-four years old, Linus Pauling created the ice-type model. Young, handsome, and by all accounts dynamic, Pauling was in his eighth year on the faculty at the California Institute of Technology and had already made several significant scientific discoveries, compelling enough that even Albert Einstein had attended one of his lectures. Still, Pauling was interested in ice; he wanted to know how ice was structured, what *exactly* it was made of, its most essential elements and its inner workings. I see a photo from that year of Pauling with a friend, sitting on a set of steps. Pauling wears a dark suit and striped tie; his arms are crossed, but he's smiling. He has grown—briefly it seems—a thick beard. His wavy hair is just beginning to recede. Pauling's friend, Norman Elliott, has his arm around Pauling, but Pauling doesn't return the favor.

It had been understood for some time by then that ice was comprised of both hydrogen and oxygen and that the oxygen atoms were fixed, in the same known positions. What wasn't known was exactly what was happening with the hydrogen atoms. "It has been generally recognized," Pauling would write in his article "The Structure and Entropy of Ice and of Other Crystals with Some Randomness of Atomic Arrangement," "since the discovery of the hydrogen bond that the unusual properties of water and ice (high melting and boiling points, low density, association, etc.) owe their existence to hydrogen bonds between water molecules." Just a few sentences later he adds, "The question now arises as to whether a given hydrogen atom is midway between the two oxygen atoms it connects or closer to one than to the other."

In what one biographer called "a flash of insight," Pauling not only realized that observations on ice's entropy meant that hydrogen atoms had to be closer to one oxygen atom than another but also that the hydrogen atoms could come in several arrangements. "Thus we assume," Pauling wrote, "that an ice crystal can exist in any one of a large number of configurations, each corresponding to certain orientations of the water molecules. The crystal can change from one configuration to another by rotation of some of the molecules or by the motion of some of the hydrogen nuclei."

I will wonder, when I read about it, where Pauling had this "flash of insight." Was he in his Caltech office—surrounded by a sea of papers—or working out a diagram before his students on a blackboard? Was he lying on a desk, as he sometimes did when he taught—"Roman-style," as he called it—or was he out walking on the Mediterranean-inspired campus? Was he away, on one of his many trips to Berkeley, to Harvard, or overseas? Was he somewhere outside, or was he in a lab, looking at ice?

Perhaps, in the end, it wasn't as important where Pauling was as much as what he saw, or foresaw. "Seeing everything," biographer Algis Valiunas writes, "was to be Pauling's specialty."

I first think about ice when I move several hours north of my parents' house for college. It's not that I hadn't encountered ice before, but carless my first year on campus, I now had to walk in ice, and great distances sometimes. I lived that year—and for all four years of school—on what was called "upper campus." My university was built on both the top and the bottom of a hill, with a major river near the base of the hill, and so there was lower campus with most of the academic buildings on the bottom of the hill, upper campus with housing on top, and then the art and music buildings were the lone part of the university sectioned across the river. The campus was gorgeous in summer and fall and even spring—tree-lined and blooming—but winter was a different story.

There were three ways down and up the almost-hundred-foot hill between upper and lower campus. First, there were the back stairs, a series of wooden steps cut right through a small, heavily wooded forest and spilling out behind the academic buildings. Much closer to where I lived, there was the road and adjacent sidewalk, which climbed all one hundred of those feet in a mere eighth of a mile. Finally, there were the front stairs, seven sets of eleven or twelve steep concrete steps heading straight down the other side of the hill, this side broad and grassy. I tried all three paths, up and down, and alternated between them, joking with friends that it was a law of averages here: one trip up the hill for each late-night slice of pizza.

The first major snowstorm of my freshman year of college, I walked down the sidewalk version of the hill just behind a guy in white khakis and a light-colored jacket. We shuffled along slowly, bundled up, with backpacks on. Suddenly the guy in the white khakis hit a patch of ice on the hill and went sliding, straight down the hill, at a remarkable speed. People in front of him cleared the way as they heard a body hurtling by. He made no noise—maybe he didn't have time to; he just kept skidding down, his backpack flying behind him. Thirty or forty feet after he started sliding, he finally stopped, in a heap. As I watched, unmoved from the spot where I'd been when he started the slide, he picked himself up, his white khakis dirty and wet, and kept walking to class.

I never forgot that student's resolve, but I quickly decided to always wear dark jeans to class and asked for some metal treads for the bottoms of my boots for Christmas. I also began to watch for ice everywhere I went; it seemed vital to understand what I'd gotten myself into. There was a thin layer of ice on the parking lots seemingly all winter; there was ice hanging off the roofs of the older buildings; ice even lined the long field just outside my dorm, enough ice that eventually they sent someone over to fill it in and add goalposts: a makeshift hockey rink.

The ice I was most interested in, though—most taken with—was the first ice of the season, the ice that was barely there and then gone. Sometimes, on early mornings when it was almost winter, my friends and I would go for long runs. The sky was dark when we started—cloudy nights turning into cool mornings—and as it lightened, we'd see the first or second frost, coating windshields and then hardening the tips of the grass. If we ran far enough, we'd sometimes pass the remnants of a farm field, long rows of what was once wheat or corn, now flanked in ice. It felt like magic; one day everything was green, and then suddenly, on those early mornings, it would be white.

The study of ice began in earnest almost 175 years ago, though it's possible individual scientists studied and collected ice much before that. In his "History of Early Polar Ice Cores," Chester C. Langway Jr. notes that as early as the 1840s and up until the 1940s scientists dug deep pits in glaciers, trying to discover their thickness. Some researchers cut through the ice manually; others used chainsaws to dig these holes, holes big enough for the scientists to climb inside. One of these early ice researchers, Ernst Sorge, lived in a snow cave underground for seven months next to his fifteen-meter-deep pit of ice. In a biography of Sorge, I read that a group of men, convinced that he might run out of food, once hiked all the way to Sorge's snow cave and pit with food and a series of winter supplies. As the story goes, the snow was so deep that only two men actually made it to Sorge's site, and on the way back, neither of those men made it back alive.

Ice studies gradually moved away from hand-dug pits and toward cutting long samples, or cores, straight from the ice. In the late 1940s and early 1950s three separate excursions set out to recover and study one-hundred-meter deep ice cores: a Norwegian-British-Swedish team on Queen Maud Land, the French Polar Expedition in Greenland, and the Juneau Ice Field Research Project in Alaska. Along with collecting polar ice cores, these expeditions began to study measurements like grain size, density, and air bubbles.

According to many scientists, though, it was in 1957 that modern ice research emerged when the U.S. Army Corps of Engineers began to extract even deeper ice cores, four hundred meters deep. When they were taken out of the ice, ice cores were then shipped to "cold rooms," placed in long metal canisters, labeled, and stored on rows of metal shelving. In black-and-white photographs from the early days of storage, there are rooms full of those metal cannisters, the ends penned with details about the ice. Scientists in those photos attend to the ice in parkas and snow pants.

Scientists had declared 1957—the year glacial drilling took off—the International Geophysical Year. Following the example of earlier international polar years, sixty-seven countries worked on scientific ventures. Along with the glacial study, *Sputnik 1* was launched that year, then *Sputnik 2*, *Explorer 1*, and *Vanguard 1* all spun into space.

Linus Pauling didn't stick with the subject of ice forever. After his ice-type model, Pauling would then go on to research ionic crystal structures, molecular genetics, quantum mechanics, DNA, and vitamin C; he'd write an introduction to a chemistry textbook and win a National Medal of Science, a Presidential Medal for Merit, and two Nobel Prizes. Finally, he would add to his list of many accomplishments and goals the protesting of the atomic bomb tests, an action that would put him, at least briefly, on the House list of those supporting un-American activities.

Though Pauling moved onto other subjects and fields, what he did do was encourage his protégé and later son-in-law, Barclay Kamb, to take up ice. Kamb took Pauling's advice, quickly rising to become one of the world's leading glaciologists: presenting at the first International Symposium on the Physics of Ice in Munich in 1968; being awarded the highest honor in glaciology, the Seligman Crystal Award, in 1977; and having an ice stream near Antarctica's Ross Ice Shelf named after him in 2003. Eventually Kamb and one of his colleagues even designed a new ice core drill, one that used hot water jets to quickly melt cylinders of ice as it cored. While some ice core drills only went partway into the ice, Kamb focused on drilling to the very bottom, where ice met land.

Kamb was interested in glaciers' movement; his research took him on trips to study glaciers in Washington state and Alaska, and on over a dozen trips to study Antarctic ice. In his later research—those Antarctic trips—Kamb found deep and wide streams within the ice sheets that cover Antarctica, streams up to fifty kilometers wide and five hundred kilometers long and that move at a hundred times the speed of normal ice sheets. "The question is," Kamb once posed, "what will happen to ice streams in the future. Will they cause a big enough effect on the flow of the ice sheet to contribute appreciably to sea level rise?" He and his own team of students and colleagues did their best to find out, collecting ice cores from some spots on the ice sheets, gathering remote video sensing data and temperature readings from others.

When I begin to look, I'm able to find various photographs online of Barclay Kamb on research trips to glaciers. In one photograph, Kamb—wiry, with thick gray hair—stands in the back of a team of researchers, leaning against a green tent in a hot water drilling camp. For some reason, most of the others in the photo are wearing Dr. Seuss hats or Dr. Seuss–inspired clothing. Kamb is in regular clothes: a long-sleeved blue shirt, sunglasses tethered around his neck; he holds a small glass in one hand and smiles slightly. In another photo, a much younger Kamb stands inside a hollowed-out piece of ice, his helmet and shoulders covered in snow. In another, he stands over the top of a narrow crevasse, skis on, poles just behind him in the snow, lining up a camera.

Whatever he was doing, Barclay Kamb always made sure to remain close to his father-in-law and mentor, Linus Pauling.

Today the largest depository of ice outside of nature is in Lakewood, Colorado. There the U.S. National Ice Core Laboratory stores and studies its own long, cylindrical glacial ice cores, currently over

17,000 meters of ice. Like the cores in those original cold rooms, the ice is stored in metal canisters and kept at −36 degrees Celsius. Also like those earlier cold rooms, the ice samples are cored from glaciers around the world and shipped or brought by scientists to the lab, who still study the ice in hats, gloves, and parkas. They saw and plane small pieces of ice off larger chunks and send these back to their own universities or research labs for study.

The basic idea of the National Ice Core Lab is both safekeeping and science; in taking samples of ice from around the world, the lab collects for research and study, but it also guards against the loss of the fragile evidence. The cores are studied; particles in the ice, air bubbles, dust, and pollen trapped inside tell the scientists studying them about changes in climate and historical events: volcanic eruptions, warm and cold spells, and increases in greenhouse gas. The ice cores are kept safe—or relatively safe—if safe means frozen.

I first hear about the National Ice Core Lab in Colorado from another teacher at my university. Between classes one day—students filing out of his course and into mine—we talk about the writing we're both doing. When I mention that I'm interested in maybe doing some research on glaciers, he gets excited. "You have to talk to my friend," he says. He tells me this friend once worked at the National Ice Core Lab, before it was housed in Denver, but he felt that there was too much government meddling with his work and finally left with his graduate students and his ice for Europe.

I imagine, while the other teacher talks, a midnight flight across the ocean and a plane packed—instead of with passengers—with large blocks of ice, taking up the seats, filling the cargo area, the pilots and stewards walking the aisles in snowsuits or parkas, a thin mist of frost covering everything.

By the time I finally do begin writing on ice, the other teacher and I no longer work at the same school. The note where I put his contact information is long gone—perhaps lost in my latest move—and I'm unable to track down his last name from any of my friends or former colleagues.

What I do is try to get a media pass to visit the lab. I will be in Denver for a conference: it seems perfect, but when I email the lab to ask, four or five weeks before my proposed visit, the person who replies tells me that there is a waiting list of at least three to four months.

Instead, from my own house I watch video footage of scientists preparing the cores for the lab, the mechanized augers twisting and cutting round pillars of ice, pulling them straight out of the snow-covered ground. I look through the lab's photographs of rooms filled with shelves of long metal canisters, each canister labeled, in black marker, with serial numbers of some sort, perhaps length and width or coordinates of the ice. I look and I think about all that ice, all those hydrogen atoms; I wonder how many configurations a hallway of ice cores might have.

Sometime after I come back from the conference in Denver, the small city that I live in has the biggest ice storm since I moved there, one of the biggest storms in its history. I watch the news from my warm upstairs apartment, above a doctor's office. Within just a few hours, more than two hundred crashes are reported: cars sliding into ditches, cars sliding into one another, a semi forking, pileups on the highway. One friend gets her car out of the driveway but then spins out of her lane on her way down one of the hills in town, her kids in the backseat; others simply abandon their cars on the sides of major city roads and trek out by foot. I decide not to chance it with my own lightweight car, one whose front driver's side window has been stuck, two inches rolled down, for months. I walk out to a nearby store and get a few groceries and then call to cancel my plans with friends and tuck in for the night. After the ice comes snow, falling steadily for hours.

The next morning I take out the snowshoes I've never once used since moving from Wyoming, bundle up in fleece pants, a sweatshirt, and jacket, and head outside. Everything has changed; the roads and sidewalks and yards are all a single flat expanse of snow. Buildings are covered in white, their roofs piled high with snow; the drifts come up several steps onto the outdoor staircase of my own red-brick building. Boundaries between one thing and the next—driveways and yards, trees and bushes, even between houses far in the distance—have become porous. It was as if I'd left out my usual door, then slipped into somewhere different, someplace more northern and less peopled. In fact, there were no people, not that I could see: no one driving cars, no one out shoveling, no one standing and looking over the balconies, only snow and more snow and ice.

I move out, snowshoeing up a single large hill and then down what might or might not be the road. There's a slight wind, but otherwise the only sound is my metal snowshoes landing softly on snow each time I take a step. I head away from the nearby business district and toward a rural road I sometimes run down. On this road I do see someone else; he's skiing in the opposite direction of me. He takes a wide angle, nodding as he passes, and then gives me my distance. I plow on, finally circling back when my face begins to feel chapped and my fingertips numb.

I end where I started, just outside the steps to my apartment.

It's there, when I'm taking off my snowshoes, that I notice: the three evergreens near the bottom of the steps are covered in ice and leaning: once-crisp individual needles look like they're encased in glass, and then that glass is covered in a thin, powdery coating. It's like those individual blades of frosted grass I saw on my runs in college: recognizable but also strange in their newness, in their ability to be changed. I walk closer and guide my hand across one of the branches; it's stiff, as I'd imagine, but the ice is thick enough that I can skate my fingers across it, that I can hold onto it like I might a fine piece of crystal, careful so it does not snap. I crouch down after a minute and look between the branches; there's a surge of snow there too, but otherwise, from that vantage point, all I see is ice, common and sublime.

Was it for Linus Pauling, I wonder, all clear and calculated—his now-proven theories on hydrogen atoms and the structure of ice— or was there a moment for him too of the sublime? Was there a line or meter, a rhythm to the arrangement of those atoms? Was the study of ice, for him, a sort of poetry?

The thing I haven't told you is this: on January 30, 1960, Linus Pauling—that great scientist and creator of the ice-type model— disappeared while out on a winter hike.

He had started on a walk near his cabin, midway between Monterey and San Luis Obispo; I see an inset map of this in an archived newspaper article: a black-and-white cutout of California with "Pauling Ranch" labeled in a rectangular box, just as large as the label for San Francisco. The Pauling Ranch—or Deer Flat Ranch, as the two-room cabin and the surrounding land was called—was adjacent to Los Padres National Forest and within striking distance of the great expanse of the Pacific Ocean. That morning, Pauling told his wife, Ava Helen, that he was going to check the fence lines on their property; he set out just before lunchtime on a deer path not far from the house. Ava Helen watched him leave.

By noon Pauling had lost his way as he walked along the cliff-lined shore. Perhaps it looked familiar, perhaps he thought it would lead him back, back to the small cabin where his wife was waiting. It didn't; instead it led up. As one newspaper writer narrated, Pauling "climbed—clawed, actually—until he found himself trapped under a large overhanging rock about 300 feet above the water. The surface there was chiefly blue shale, slippery and dangerous." After getting up on the cliff—an eighty-degree ledge—Pauling could not get down. He was worried that any movement he might make would toss him into the ocean below, and so there, in a light sport coat and slacks, he waited.

As the night went on—and a false report began circulating in the news that the great scientist Linus Pauling had been found dead at the bottom of a cliff—Pauling followed the movement of the constellations to track passing hours; he recited to himself the periodic table; he counted as high as he could in several languages. Finally, out there in the dark—after a search party called his name but could not hear him calling back, after his wife and his daughter and his son-in-law Barclay Kamb began to become extremely frightened—Linus Pauling covered himself with an unfolded map and "lectured the surf," as one newspaper reporter would put it, on all he'd long ago discovered about the nature of chemical bonds.

# THE ICEBERG PROPOSAL

In 2017 the National Advisor Bureau, a private company from Abu Dhabi, proposed solving the United Arab Emirates' freshwater shortage by using powerful boats to tow a three-kilometer-long iceberg from Antarctica over 9,000 kilometers across the Indian Ocean to the eastern coast of the UAE. After the yearlong tow, the ice would be sliced into small blocks and then those blocks melted. The project's directors argued that a single towed iceberg could supply the entire population of the country's capital with drinking water for about eight years.

I learn about the plan—the Emirates Iceberg Project—from a website, *Science Alert*, that the same day also features the articles "The Mystery of This Tiny 'Alien' Skeleton Has Finally Been Solved" and "Massive Oil Fields in Texas Are Heaving, Sinking, and Opening Up Like Mouths." Skeptical, I probe further. I quickly find out that indeed news of this project is all over. By the time I check, the *Guardian*, the *Independent*, the *New York Post*, *Newsweek*, and the *Huffington Post* have all picked up the story. "The icebergs are just floating in the Indian Ocean," the director of the Emirates Iceberg Project says in one of these interviews about the project and its plans. "They are up for grabs to whoever can take them."

## The Iceberg Proposal

Several of the articles link to videos the Emirates Iceberg Project had uploaded to YouTube. One begins with the animated characters of a young boy and a man, standing alone in a desert, holding a book entitled *Filling the Empty Quarter*. Soon the video pans out to images of icebergs, bright white and looming massive above an aqua-green ocean, and against black and then blue skies. "Climate Change Is Causing Ice to Melt around the World" the subtitles begin, "Wasting Billions of Gallons of Fresh Water on Earth." The video pans over oceans and huge cloud-covered mountains, moving from animated images to real ones. One frame shows a tall blue wall of a glacier calving directly into the sea, snow rising in a mass from the water; another frame shows a second glacier calving, this time small pieces of snow falling first, before a colossal wall of snow drops, as if in a plummeting elevator, and splits into the sea. "Ice Is the Purest Source of Water Known," the subtitles continue. "That Source of Fresh Water Can Be Brought to the U.A.E. from an Island Called the Heard Island Which Is 9,200 Kilometers Away."

The video moves on to show an animated image of a beach, umbrellas and palm trees waving in the wind. Just beyond the beach is a giant block of ice: an Antarctic iceberg, covered in penguins and polar bears.

The National Advisor Bureau isn't the first organization or even person to suggest towing icebergs. In 1949 oceanographer John Isaacs—now recognized as the "godfather of the modern iceberg towing movement"—gave a lecture at California's Scripps Institute of Oceanography where he proposed towing a glacier from the Antarctic to San Clemente Island, this time to solve California's water problems.

The proposal had evolved for Isaacs over time. According to his biographer, Daniel Behrman, Isaacs came upon the idea after wondering whether it would be feasible to create a pipeline connecting the Columbia River and southern California. "When I started to optimize it," Isaacs relayed, "I saw that as you make the pipe bigger, the cost of moving an acre-foot of water becomes cheaper. Obviously, I should have known at the time that it would never stop optimizing." Soon Isaacs realized it would make sense to, as he put it, "shorten the pipe and just use it as a towed container." As he began to think about how a massive container could be built to hold all that water, he realized something new and surprising. "When you optimize your container, you see that you have come just to the dimensions of the ordinary tabular Antarctic iceberg, a free package. You begin to see how little energy per acre-foot is needed to move it and you realize you might as well start in Antarctica."

Isaacs's was one of a series of seminars at the Scripps Institute. Searching through the University of California San Diego's special collections, I see a black-and-white photo from one of these seminars, maybe the first lecture that Isaacs gave or maybe another. There are three or four women and about thirty men sitting in small, right-handed desks; some are in suits and ties, some are in lab coats, all facing forward. Two windows are open; the rest have the shades pulled. No one in the room is smiling, or not quite, but they seem to be rapt, listening with marked interest to whatever Isaacs is saying.

One winter, a couple of years before the Emirates Iceberg Project is proposed, I have a chance to visit the Scripps Institute myself.

# The Iceberg Proposal

I'm in California with my family over a long weekend. It's March when we travel to San Diego, and my parents have rented a guesthouse just off the beach, close enough that on early morning walks we can see a cove filled with forty or fifty seals sunning themselves, flopping in and out of the water, and we can watch the reflection of the sunset over the waves even from the living room window.

The Scripps Institute isn't far from where we're staying—maybe a ten-minute drive at most—and so one afternoon I steer our rental car up a set of hilly roads, weaving toward the institute's drive. We miss the turn, the small sign for it receding behind bushes and flowers on one side of an intersection. On the second pass we see the sign and follow the scrub- and pine-sided road a bit farther. Just outside the entrance to the Birch Aquarium—the institute's showcase for the public—are two circular fountains. The first features two life-sized bronze whales diving out of it, and the second, just to the side, features another whale's tail. We pass by both and into the building.

I don't see anything about towing icebergs in the institute's public collections, but there is an exhibit entitled *Feeling the Heat*. After wandering through an aquarium that houses California moray, sheep crab, and red rock shrimp, I walk around the exhibit, through the artifacts, displays, and even hands-on activities related to global warming and oceans. I watch films and see signs and diagrams about carbon and marine life and rain. "As you warm the upper layers of the ocean," one video explains, "that warm water is lighter, and it is much harder now to be able to mix the ocean vertically...we put a lid on the ocean."

There's a wall-sized graph of $CO_2$ in the atmosphere too, with the question "How high will it go?" in black letters on yellow paint, pointing toward the graphed $CO_2$ line. The line rises and drops in fits and starts along a horizontal plane through various ice ages. It becomes nearly vertical in showing the increase in carbon parts per million from the Industrial Revolution until today.

It wasn't until fairly recently that ships—and the companies that employ them—began thinking of traveling to icebergs. For most of shipping history, avoiding icebergs—not trying to hunt them down—has been the goal.

Before 1912, the year the *Titanic* sank after colliding with an iceberg near the Grand Banks of Newfoundland, ships mostly relied on other ships' radio calls and on their own lookouts, usually supplied with binoculars, who took shifts in the crows' nests to watch for other ships, icebergs, or land. After the *Titanic*, some big changes took place in the realm of shipping. Only one year later, the first "Safety of Life at Seas" conference was called between world shipping powers. I see the original minutes for one part of this conference. The committee recording of the minutes is originally typed out, in all caps, "COMMITTEE ON SAFETY OF NAVIGATION." In pencil, though, "ON SAFETY OF NAVIGATION" is crossed out and "ON LIFE SAVING APPLIANCES" is added just below the original title, also in block letters.

The conference did call for, among other things, lifesaving appliances like the standardization of lifeboats on ships and also increasing regulations that all ships' watches must be twenty-four hours. At the first conference there were also appeals for regular and regulated ice patrols. Soon the International Ice Patrol was founded with the goal of monitoring icebergs in the North Atlantic during the high shipping season of March to August, sending out twice-daily radio calls about these icebergs to ships.

Gradually, technology improved and there were more ways to monitor icebergs. Both radar (electromagnetic sensors) and sonar (equipment that can hear sound waves reflected from objects) began to be used more widely in the early 1900s, allowing ships to detect icebergs even a few kilometers away. More recently, the International Ice Patrol and others have turned to computer plots, satellite imagery, and real-time maps to notify ships of the whereabouts of icebergs.

The system is much improved, but it isn't perfect. According to the Institute of Ocean Technology, between 1980 and 2005 there were 2.3 iceberg crashes a year. Even with all the modern technology, there's still a sense, it seems, that we need to be vigilant to track things above and things below.

I learned the same basic idea of paying attention to both what's above and what's below when my brother taught me to sail, although it wasn't icebergs then, just underbrush at the bottom of the lake. I was ten or twelve at the time, and my brother two years older; we were on a two-person sailboat, coasting across a small lake in northern Wisconsin, a place we came to every summer. He'd taught me the fundamentals of sailing the year before: where to sit, how to raise the white fabric sail, how to harness the wind, and how to shift the rudder in order to tack, moving our narrow boat from one part of the lake to another. It was perhaps in our genes; when my father was about our age, he and his friends used to sail in the early mornings on another Wisconsin lake. One of his brothers even built his own boat, a small pram that they'd take out in the harbor.

This particular summer afternoon, as we were coasting toward shore, the sail curved, the rope flapping against the mast. We'd made it almost all the way across the lake, straight to the densely wooded shore, when suddenly my brother turned the boat fast, too fast, and we were tossed out as the mast tipped down and the boat lurched over, rolling into the water.

I'd learned—again from my brother—how to right a boat when it tipped. For several minutes, treading water beside the upside-down boat, we tried, moving onto the centerboard and leaning back, tugging at the boat's sides. But this day, we couldn't upright the boat. As hard as we pulled, it wouldn't move. Finally, I loosened the straps of my life jacket and ducked my head underwater to see what I could see. It was dim and cloudy, but I could see the mast facing straight down, a foot or more of its tip stuck in the shallower-than-we'd-imagined lake-bottom muck and weeds.

When I ask my brother about this story some years later, when we're in our twenties, he remembers the story differently, with my action making the sailboat tip, not his, my inexperience, or perhaps my shifting weight throwing us off-kilter and the boat into the weeds. I will wonder for some time after if it mattered whose fault it really was, or if perhaps the crash was inevitable, two young kids only skimming across the surface.

According to the *Oxford Dictionary of Word Origins*, the first recorded use of the phrase "tip of the iceberg" was in the 1950s. The *Online Etymology Dictionary* puts the date a bit later and more exactly, at 1962, although neither source adds any details about who first mentioned the phrase or where.

The idea of using an iceberg to talk about literature is much more traceable. In the 1920s Ernest Hemingway began to talk about "iceberg theory" or the idea that an omission can strengthen a story, leaving what was unsaid up to the reader to understand, through images or subtle phrasings that suggest something important is happening just below the surface. It was an idea that Hemingway had taken from his time working as a journalist, when he only had space on the page for the immediate and was unable to include even relevant, seemingly helpful background. When he began writing fiction, he tried the same approach.

In *Death in the Afternoon* (1932), he wrote: "If a writer of prose knows enough of what he is writing about he may omit things that he knows and the reader, if the writer is writing truly enough, will have a feeling of those things as strongly as though the writer had stated them. The dignity of movement of an ice-berg is due to only one-eighth of it being above water. A writer who omits things because he does not know them only makes hollow places in his writing."

The first time I hear of Hemingway's iceberg theory, it's in the context of subtextual dialogue, of saying more than what you are saying. My friend and I who are talking about the idea speak for a while about the risks: being opaque, elusive, confusing. What we don't talk about, though, is the equal possibility of discovery, of finding that the thing just below surface is more wonderful or more terrifying than we ever imagined.

In actuality, the tip of an iceberg can vary widely. Tabular icebergs, the kind featured in the Emirates Iceberg Project's video, are generally flat on the top, with nearly vertical sides above the water. Nontabular icebergs—a catchall designation for every other type of iceberg—come in a variety of shapes, from rounded domes to ice wedges to the mountainlike pinnacles that are commonly shown in photographs and illustrations. The tallest icebergs have been nearly 500 feet high, but icebergs can be much smaller too, technically starting—according to the National Ocean Service—at dimensions of 16 feet high and 98 to 164 feet thick and covering at least 5,382 square feet. In any case, icebergs all begin by breaking off of larger pieces of ice, whether glaciers or ice shelves on land or other larger icebergs in the water. When those icebergs break off, they float in the denser saltwater ocean because they're made of less-dense fresh water and are also part air, part water.

What starts as the tip of an iceberg doesn't always stay as the tip. Even though more of an iceberg is under than above water, they have the propensity to "roll" or flip sides from top-to-bottom, especially when calving or when parts of the ice begin to melt. Birds, some sources say, are known to take flight just before the motion begins. They can sense, it seems, that something big is beginning to happen; they feel the ever-present risk that what's below will soon be on the surface of things.

It is a risk, and not just for birds but for boats and for other animals, fish, and plants living in the ocean, and even for nearby land. In a 2012 study, physicist J. C. Burton and a team of researchers set out to understand the effects of flipping icebergs on the ocean. Working with plastic models that replicated the weight and density of real icebergs but that could be studied in a laboratory water tank, Burton and his team studied the waves once the models flipped. A rolling one-kilometer iceberg, they learned, can cause land to shake as if there were an earthquake. The iceberg's motion releases the same amount of energy as an atomic bomb.

I see an image of one of these upside-down icebergs in *Smithsonian* magazine. The photo was taken by filmmaker Alex Cornell, who was vacationing with his family in Antarctica. The iceberg is blue, gray, and black, and its glassy surfaces rise toward a small pinnacle but reflect dark in the water, against a backdrop of nearly white mountains and wholly white sky.

John Isaacs first gave up on his idea of towing icebergs because he didn't know how a boat might attach lines to the ice or where the underwater tow would go. Six or seven years after his seminar at Scripps, however, he returned to the idea, practicing on six-hundred-pound blocks of ice and suggesting, after finding out that small icebergs had once been taken to Peru by adding sails, that sails might be added to larger icebergs too, to help guide their journey. Isaacs estimated that it would take about two hundred days and cost about a million dollars, with the first five days spent simply getting the iceberg moving at half a knot.

Isaacs's wasn't the only towing plan. In his 2011 *Atlantic* piece "The Many Failures and Few Successes of Zany Iceberg Towing Schemes," Alexis Madrigal details a range of iceberg towing schemes from 1825 onward, including plans to convey "by means of pipes and air-pumps, the sea breeze to London," plans to fit a large screw through an iceberg, and plans to wrap an iceberg "in sail cloth and plastic" and tug it behind a ship.

In the Emirates Iceberg Project video, two large boats sail around the squared edges of an animated iceberg. Attached to the backs of the boats is what appears to be a giant net, maybe metal, that looks like a fence. The white boats circle closer and closer until the iceberg is surrounded with the fenced material. With a giant rope attached to the fencing, one of the boats begins to pull and the animated iceberg begins moving through the water.

The thing is, towing an iceberg isn't as simple as it sounds. Or perhaps it is as simple as it sounds, which is to say not at all. Icebergs are unwieldy, generally 80 percent or more underwater. To make the cost and trip worth it, the icebergs Isaacs and the Emirates Iceberg Project proposed towing would have to be huge, several kilometers in size. Of course, there are problems beyond even the basic concerns of feasibility; all these plans raise important questions about the environmental impact in Antarctica, economic cost—between $100 million and $500 million, many estimate—and the effect on plant and animal life caught up en route. And there's the question too of whether the icebergs would even make it across the water before breaking up in the Southern Ocean's strong undercurrents.

Along with aquariums and educational displays, the Scripps Institute displays photos of glacial ice: the Upper Glacier in Argentina, the Muir Glacier in Alaska, the Larsen Ice Shelf at the South Pole, the Franz Josef Glacier in New Zealand, and the Pasterz Glacier in Austria. The first photo, of the Upper Glacier, was taken in 1928. It's a black-and-white photo with two snow-capped mountains in the distance and a field of white ice in the foreground. Just below that photo is another shot taken from the same angle. The mountains are still there, but in the place where there was once ice, there's only water. The photos of Alaska's Muir Glacier—dated 1941 and 2002—show the same change: one photo with ice, another with water. In New Zealand's Franz Josef Glacier, there's ice in both the 1867 and the 2002 photos; the difference is just the amount of ice. In the older photo, the ice stands imposing, a massive wall in front of a mountain range; in the newer photo, the ice is present only in a valley between the mountains. The Austrian Pasterz Glacier is present in the 1850 photo and completely gone in the 2004 photo.

Photos of the North Pole and the South Pole are taken from space. In the North Pole shot, the ice that used to touch Asia is now far from it. In the South Pole photo, the Larsen C Ice Shelf has changed from a fairly solid line of white to hundreds, maybe thousands, of fragments—icebergs—floating away from the continent.

One photo, the largest in the display, has no precursor. Dated 2005 and labeled "Greenland River of Ice Melt," the image shows five or six people in the distance, on the snowy bank of what appears to be a winding river. There are low clouds, but otherwise the entire back of the frame is a wide expanse of ice, white in some spots, brown in others. Where the people stand, the river is perhaps five or six feet across. By the time it hits the bottom edge of the frame, it's widened to ten or twelve feet, maybe more. The water is sun-streaked near the people, but it grows darker in front of them as the walls of ice around it grow higher and the water plummets between them.

I study these photos for some time. Eventually I keep walking and turn to see what else the institute holds. Not far from the glacier photos is another display on carbon dioxide emissions. I almost pass it by before I notice its title: "The Human Volcano."

The first night we're in California, my father and I wander to the beach. My mother has stayed back at the cottage we're renting, reading a book by the fireplace. It's early evening, just before sunset. We're staying only two blocks from the ocean but still aren't sure the nearest path to get to it. We cross a busy street and wander through a narrow lane filled with houses and then down a steeply pitched road. As if on purpose—just the right place, just the right time, in that exact space—there are wooden steps at the end of the road, maybe public, maybe private, we don't know, but they lead straight onto the sand. My father and I follow them.

The beach is a mix of sand and rocks. The sand is wet and smooth from the incoming tide. The rocks too are smooth and dark, some covered in bright green layers of algae, others higher and dry. A name is carved into one of the rocks, though the first few letters are too washed out to make out. Just above the name is a sketched numeral thirteen.

My dad and I walk for a while until we find our way to a broad flat stone. The water laps up, almost onto it. Lines of white waves crest not far off in the distance. Two young boys appear to be learning to surf nearby, but otherwise it's quiet. Narrow clouds broken up by a scattering of white and yellow sky meet the waves. In the photograph I take of this scene, my thinning father will become a dim outline, backlit by all that sky. "Dad, San Diego" I'll write below the photograph when I get home.

I don't remember what my thoughts were when I took that photograph or even when I looked out over the Pacific Ocean. I'm certain I did not think of icebergs that day, did not imagine giant salt-splashed ships in the distance, moving in from the Arctic or the Antarctic, towing on the water toward us mountains or hills or just blocks of ice—or what remained of the ice, after miles at sea—pinned or fenced or wrapped or pulled behind them. I did not imagine ships colliding with ice, tipping in one quick, or maybe impossibly slow, motion, splitting at the hull and sinking into the sea. As I looked out at the waves I didn't wonder what sonar sounds like, whether, as I will read in an encyclopedia entry sometime later, it "squeals, pops, and creaks" from all that cracking ice.

Can we grieve, I wonder now, what we have not yet lost, those losses that are still to come? Are they just under the surface waiting to be struck, or do they stretch out before us, early evening tides, dispatches landing on our shores?

There was no ice off the coast that day in California, only my father and I and the fading light, rippling on the surface of water.

In 2003—fourteen years before the news about the Emirates Iceberg Project—the *Smithsonian* profiles the *Norseman*, a 4,600-ton ship, and its captain, Jerome Baker. The ship is 270 feet long and has 9,600 horsepower. Baker and the *Norseman* run the seas near St. John's, Newfoundland, for a month at a time. Some of Baker's work involves ferrying gear and supplies. The rest of the time he tows icebergs out of the way of a North Atlantic oil company.

"It's no big deal," he says about spotting icebergs in his path; "it's all part of the job."

A couple of years later, thinking back on that trip to California and John Isaacs's iceberg proposal, I will open a map on my computer to try to calculate the route an iceberg might take, or the route Isaacs might have imagined that it would. "Sorry," the mapping search tells me in response to my query, "we could not calculate directions from 'Antarctica' to 'San Clemente, California.'"

# THE SPEED OF FALLING

On the simplest level, the speed of falling is the acceleration of gravity times time, or velocity = $9.81 \text{m/s}^2 \times$ time.

I imagine it went something like this: he was running, bounding up the trail, two steps at a time, one leg moving forward while the other was finding its footing, getting into place. I imagine it was perpetual motion, his hips flexing, his knees rotating, the long contours of his lateral muscles not once losing the tempo's pace. His steps were lanky and light, his breaths fixed and stable, his lungs forcing the air out of his diaphragm, elevating his ribs in quick one-two one-two bursts. His shoulders were square at first to the ground, but when the mountain became steeper they took a fifteen- then thirty-degree angle. His mouth was open, but he leaned into the rhythm, into the hike, into his early twenties body, a body that could do anything, that could feel nothing, nothing but the heat of movement, of hot hope on a June day. It was hot. He was hot, so hot he'd already taken off his light spring jacket; he was wearing it like a belt around his waist.

I imagine that he went fast, so fast that sweat gathered around his short, light hair in the way that it often did, beginning at his scalp and moving out, across his barely receding hairline. I imagine that water beaded on his face, on the edges of the contoured lenses of his glasses. It stung his eyes and pooled in the un-tanned creases on the backs of his legs, soaked through his shirt, left a dark cross at the intersection of his spine and his shoulder blades. I imagine that sweat dripped off him onto the dry ground, one drop at a time until there were drops trailing behind each of his quick steps, falling off his body in perfect translucent circles, landing for a moment before soaking into the grass or the dirt or the long expanses of rock. I imagine that there were hundreds of drops up that mountain and then thousands. I imagine that if we'd known soon enough to track them, they might have led us, like a constellation of tiny stars, like disappearing swimming pools.

In his unfinished work *On Motion*, Galileo argued that different weights fall at different rates. His experiments showed, he said, that lighter bodies moved ahead of heavier bodies at the beginning of a fall, but later, heavier bodies overtook the lighter ones and arrived on the ground first. Though some historians wonder if he actually performed the experiments, Galileo said he came to his conclusions after dropping equally sized iron and wooden balls off the side of the leaning tower of Pisa.

If you walk straight west from Fuglesteg—the stone house, the low-lying mountain, the path we'd been traveling that day—you would reach the very tip of the Sognefjord, the deepest and longest fjord in western Norway: our fjord, with its aqua ice-fed waters set against our small seaside village, some strawberry farms, and miles of birch- and spruce-covered mountains. You would see a long blue dock stretching into the water, a concession for the cruise ship tourists who now come five or six times a summer, and then the white three-story guesthouse with its fading black tin roof, the sort of ceiling that makes every summer storm sound like an ending.

If you traveled straight east from Fuglesteg, you would reach the widest part of the Jostedalsbreen Glacier, via a high mountain pass, covered in snow most of the year, and a rocky plain, Turtagro, where Olympic ski teams often race in blue-jacketed packs for springtime practice. To the south, you would see more small villages gradually becoming towns and then the cities of Bergen, Lillehammer, and Oslo. To the north, you would cross more mountains and more fjords, and then eventually you would find yourself at the exact point where the Aurora Borealis—in its brilliant flashing green—lights up the dark expanse of the Arctic Sea.

Of course it's impossible to walk anywhere directly through the high mountains in Norway. Maybe this is why the farm, the stone house, and the trail are all called Fuglesteg in Norwegian or, in English, "the path for the birds."

It was my suggestion that we would walk the path for the birds. We were eating breakfast in the guesthouse that morning, watching the river outside break against the banks. The sun was out; some sheets blew on the clothesline; a white curtain swayed in the wind.

It was his first summer in Norway, but it was my sixth or seventh there, hiking and kayaking and helping my friends fix up the old three-story house in the center of that tiny Norwegian farming village.

There were a dozen things we could have done that day, an uncharacteristically warm day for early June, in that place that was so urgently beautiful. One of my friends suggested Bolstad Nosse, a grueling five- or six-hour hike to the top of a local lookout point. Another mentioned a local church, suggesting maybe we could hike there or take a bus and walk the long grassy hill behind it.

How about Fuglesteg? I offered instead. *The hike would be easy but not too easy; the views are always worth it*, I reasoned. There was a stone house at the end of the trail—hundreds of rocks, placed by hand, in an exacting symmetrical formation—that would be a good place to see.

Everyone had agreed to Fuglesteg, but especially him. He always agreed, was always up for anything, outside, with friends, and especially if it would serve them, make them happy. And the day—or the beginnings of it, at least—was happy. We had walked the trail in twos and threes on an eight-inch-wide dirt path through a farm field, up a double-tread road through a cluster of white birch trees, or something that looked like them, and then onto a single rock path that switchbacked, every few hundred meters, steeply. We walked past a tangle of power lines and a pounding waterfall that we could hear before we could see, past a graying barn and a grassy knoll. At almost every turn, we could see the tree line and snow in the distance and open, expansive views of the whole valley, straight to the sea.

I learn about the iron and wooden balls that were possibly dropped off the tower of Pisa at the Museo Galileo in Florence, Italy. I'm with my friends Andrew and Laura. They've been in Italy a few weeks already on summer holiday and invite me to join them for five days. We talk about going to Rome or Venice or even Pisa but decide, in the end, to spend a few days in a quiet farmhouse in Tuscany and the rest of the time at a city-center apartment in Florence.

The heat is stifling. I feel it late at night, as soon as I walk down the metal airplane steps toward the runway and the airport doors; I feel it again when I stand outside the front of the airport, trying with little luck to catch a cab to the city; and I feel it when I wait, at the wrong train station, for my friends for over an hour. It's hot enough I sweat right through my shirt and immediately feel I have made the wrong choice in packing no dresses.

The whole time we're in Italy, the heat never lifts; everywhere we go, the Italians we meet tell us it's the hottest summer on record, that it will no doubt cool down sometime soon, that it has to. At the farm there's a pool that we swim in multiple times a day, but in Florence there's no reprieve. Within a few hours of my first full day in the city, I've badly sunburned my neck and cheeks and I've come down with a heat-induced migraine. Laura and Andrew seem better adjusted than I am, or better packed at least; they've brought matching straw hats, a large tube of sunscreen, and only lightweight clothes; compared to me, they seem to glide through the place.

"I'm melting," I tell Laura after a morning wandering through the streets, following Laura's tour book to various Renaissance statues and important pieces of architecture.

"How about a museum?" she asks and then says that while I cool down inside, it will give Andrew and her a bit of time to finish some shopping, that perhaps we can meet up later that day.

I look through the list of museums in her tour book: the Uffizi Gallery, the Accademia, the Innocenti Museum, the Pitti Palace. One stands out and is also close to where we're walking.

"I'll visit the Museo Galileo," I tell her.

Though the news that would come across televisions and computer screens that first week in Norway and then later in England and then the States would not lay blame, would call it a "tragic accident," the hike to Fuglesteg was not an accident; it was a well-planned afternoon.

The trail was busy: other groups of hikers and walkers dotted the path, but we moved along steadily, up and up the trail. We were two-thirds finished with the hike, almost to the last rocky ascent, when someone mentioned wanting to get a good spot to eat our lunches. I was too hot to hurry; I'd already stripped down, one layer after another, until I was only wearing a light shirt and shorts; the girls behind me had stopped for three or four water breaks.

He offered to run, to get there first, and to wait for us at the trailhead.

The path was dusty and some of the trees hung onto the trail, but when he started up the trail ahead of us, he was like a sprinter just off his block. By the time the rest of us got to the stone house, he had been there for some time, waiting, asking if anyone would like him to get them a drink of cold water from the freestanding spigot.

Still, even with the run, it wasn't enough: the trail we'd done, the hike, the bicycle ride he was planning to take home. He wanted to see the view from the top, to take a photograph, a memory to bring down to the rest of us and then back home for the winter days at university on the British coast, studying for his PhD in mathematics, "maths" as he called it, where he worked on not just formulas but *real-world problems*, where he tutored his friends and acquaintances in the early mornings, telling them that good mathematics *always* makes sense.

We were sitting in a circle, eating cheese and jam sandwiches, when he turned to me and said, "I think I'll go on a bit farther."

Galileo was in good company with his ideas in *On Motion* about different weights falling at different rates. The theory made sense. It was Aristotle anyway who first argued that objects fall in relation to their weight. This was because *all motion*, Aristotle had reasoned, was, at the crux, a *product of change*.

I imagine that he followed the trail around a bend and then upward, as far as he could go. Though he did not tell us that he was going to the high ridge above the house and the valley—the climb's farthest point—and maybe didn't even realize this was the endpoint he had in mind himself, perhaps his body took him there, and maybe always was taking him there. I imagine that he could see the stone hut: small, rectangular, each stone of a different size and height, but looking sheer in the distance; there, on the edges of the forest were the brightly colored shirts of the rest of us, making our way down the mountain in slow file. It was quiet, except for the crunch of needles underfoot and leaves. It smelled of pine, like the long rectangular table where we ate bread and jam for breakfast, like the exposed ceiling beams in the top floor of the house, like the forests he had walked for several years with his family, on the coasts of England.

I imagine that's when it happened. Maybe the stone house was obscured. Maybe the path narrowed but he didn't care. He was still moving too fast to think, to stop, almost to breathe. His legs propelled him forward, always forward. The ridge was just ahead, so close he could see himself there at its exact crest, with a perfectly proportioned cirrus sky above and beyond, all around his broad body. There was a road on the other side of the ridge, one that led straight to the glacier and to the mountains. He thought that he saw a glint of sun on the blacktop.

Maybe he turned, maybe he pivoted or tried to get a better view of the road or the stone house or the rest of us, and that was when it happened; he stepped back and his foot caught nothing. And then at eye-level he saw the tree line and the granite side of the mountain and he felt the air lift his shirt.

I imagine that his phrase matters, "go on a bit farther," instead of "go on farther" or "go a bit farther" or simply "go on." I imagine that every detail matters: the brand of shoes he has worn, whether his camera takes AA batteries, the size of the water bottle that he has brought with him and the way he filled it up halfway, how he has eaten not one but two sandwiches, the fact—I realize some hours later—that he has taken off his jacket, that he has tied it around his waist.

It will not matter that he has taken off his jacket; his jacket was dark; his shirt was dark; his pants were dark. *Do not wear dark clothes*, a man from the rescue team will tell us when he asks for a photograph for reference from one of our cameras, on the first day, in the first hours, when it seems he still might be found, and then again some days later when we are standing on the same overgrown lawn between our house and the rest of the village, after a policeman dressed in navy blue with a white stripe on his jacket and his pants has come to the door holding his hat in his hand, after we have left full plates of food on the table uneaten for two days, after the coroner has declared that she cannot officially identify the body found in the woods as him because every one of the teeth is broken from the fall, from the *five-hundred-meter dive into the woods.*

Except it was not a dive, nor was it a leap into the black. This one thing I want to tell; this I know.

It's cool in the museum; I feel it as soon as I enter, a wave of cooler—though still not cold—air. I buy a ticket, follow the signs to the Medici and Lorraine collections. I see telescopes, thermometers, giant globes, discussions of the history of science and of electricity. I take photographs of various artifacts in each of the collections. I stand from several angles to capture on film the largest of the golden celestial globes.

Once I've made it through several of the exhibits, I sit on a bench in the corner of one of the upstairs rooms to rest for a little while; it's one of the few benches I've seen, in the museum or the city. Families and couples and individuals walk by with guidebooks or museum brochures, talking softly. A few kids sprint past, their mother following and calling something at them in Italian. A museum guard circles by a few times. On the third time, I wonder if he's going to ask me to move along, but he doesn't.

I notice after a while that most of the people in the gallery just ahead of where I sit seem to be stopping and gazing at something in a glass case. Everyone who moves into the room stops there and stays, pointing or taking photographs or simply circling around it.

I get off my bench and walk closer. The glass case is shaped to be a perfect, translucent egg, laced with a single band of gold. It stands atop a narrow ivory column, also gold-plated. Even still, I can't tell what's in that case at first. It's not until I walk to the very edge, so close that I could reach out and touch the egg if I wanted, that I see what's inside. Browned from time, but arched straight up to the sky, is one of Galileo's fingers.

Fuglesteg was the second stone house. The first, built in the early 1800s by two brothers who stacked stone on stone without even telling their wives, had burned in 1985, from the rafters out.

I learn this from some friends in the village, Edvin and Brit. They're over at the house one day, helping us find tools and ladders to repair the siding on the guesthouse, and I ask Edvin about the history of the stone house and how it got there.

"Well, that was the second house," he says, "the one up there now. Come by our house; you can see old photographs in our town history book. I'll tell you the rest of the story."

I do stop by Edvin's house, one late afternoon. I knock on the door to see both Edvin and Brit standing there, their shoes and raincoats on, ready to leave. They invite me in anyway, and we stand around their kitchen table looking at a book of history from the local area with photographs of burned-out rafters and, later, people coming together to resurrect the house. In one photograph, only a small stack of stones remains, lining the foundation. In another, white dots of snow or rain obscure the foreground of the frame, and in the background, people are building.

The day of the hike, we almost didn't make it to Fuglesteg. That day we drove right past the trailhead, all the way across the ridge. We didn't mean to cross the ridge or even to come anywhere near it; we turned the wrong way at first, though, trying to find a parking area, and headed up the two-way, one-lane mountain road instead of down it. Our driver—the owner of the car—was reticent to attempt a turnaround in too narrow or too steep a space, and so we continued driving higher for several minutes, curving our way around the contours of the mountain.

One of the guys, Tom, was following us that day on a bike, unaware that we had missed the turn and that he did not need to ascend three or four hundred meters. We laughed as we watched his legs move rapid-fire as he stood on his pedals, the thirty-year-old metal bike frame shifting left and right. We yelled out the window, *Go down; wrong way; turn around; stop biking*, but he did not hear us against the wind and the noise from his own hot breath and the car, and so instead he kept pace with our own motion, thirty to forty feet behind.

When we finally found a wide enough swath of road to turn the six-seat station wagon around, we had almost made it all the way to the top of the mountain. There was snow everywhere, but the light coming through the open car windows onto our faces was sharp and warm. Tom was there too, just behind the car, never once complaining about the missed turn or the long way up.

"Who would ever believe," Galileo wrote in *On Motion*, "for example that if two lead balls were dropped from the orb of the moon, one a hundred times larger than the other, if the larger reached Earth in one hour, the smaller would take a hundred hours in its motion? Or also, if from a high tower two stones, one stone twice the size of the other, were flung simultaneously, that when the smaller was halfway down the tower, the larger would have already reached the ground?"

Eventually Galileo came to recant his earlier theories on velocity. He would say that his first impulse had actually been to discount weight, that he'd come to his incorrect conclusions through experience rather than logic, and that, upon standing outside in a hailstorm, he realized that hailstones of different sizes—"different bodies," as he called them—were all traveling toward the ground at exactly the same speed.

Maybe he didn't feel hot and so he didn't stop or pivot to look at the white stone house or the rest of us in our brightly colored clothes. Maybe he did make it to the top of the ridge. Maybe he had already been there, and it was when he came down that he started to feel lightheaded, like he had stood up too quickly in the shower, like the corners of his eyes were beginning to go dim. Maybe it was then that he realized he had moved too fast for his own good and so he stopped, wisely, to drink some water, to get his bearings, to make sure that it wouldn't be the first time his legs gave out.

Maybe he leaned on a tree for a moment; maybe it was a sapling and he knew that it could not bear his weight, but it felt good to rest his back upon something, to feel—however faintly—that his body and the mountain were interconnected, two sides of the same formula. Maybe there was a spot taking over his vision on one side of his right eye. It might have been the summer sun; he didn't know and so he crouched, got low to the ground. His calves burned, digging into the steep incline. Maybe he told himself that it was worth it, that it was always worth it: the backcountry, the mountains, the lake district where he'd gone with his father and his friends, the marathon he ran only a month before, no training, just grit. Maybe he reached into the side of his backpack for his water and took a long drink. Maybe the cold felt good on his larynx. Maybe the liquid went down easy. Maybe he looked ahead and behind him.

In this version of the story, this was when it happened. It was when he bent down, open backpack in his hand. The ground beneath him was dry, but he did not realize it was not solid. A small piece of dirt gave way under the shifting of his legs and then suddenly a larger piece split off and tumbled below him. He lurched off his tree, off his footing, off the steep slope of dirt and brush and too-thin trees. He reached forward with his open arm, flailing for a branch, some grass, anything that he might hold onto, but it was all just out of reach.

There it was: the granite side of the mountain and the tall trees and the open air all around him.

Or maybe he blacked out. Maybe he tripped. Maybe his backpack released from his hand as he rolled and dipped. Maybe his camera dropped and he staggered to get it. Maybe it flashed through his mind that the camera was borrowed—his father's—like his frame, his urge toward kindness and courage, toward keeping his promises. Maybe he made promises. Maybe he knew there was no time for promises. Maybe he yelled but no one heard him or someone did hear him but couldn't recognize the voice or the person or the sound. Maybe he saw that between him and the ground was 500 meters, 1,500 feet, 18,000 inches of air. Maybe he did the math. Maybe he knew that gravity times time would be about 5.6 seconds. Maybe he realized that it was all breath, all weight.

*Grief is the cost of love.* This is what a psychologist friend who comes to the guesthouse after the accident says. We are sitting in the river room. It's a pale yellow— three windowed walls ready to hold the light at any angle—I painted it myself some years back, from a white faded to brown after years of neglect. Couches and chairs circle the outside walls; footrests and casual coffee tables drift together in the center. There are candles hanging from the ceiling in mason jars beside stacks of old books. You can see the mountains, the river, and the fjord from any position in the room. This is why we call it the river room, after the Norwegian name for the guesthouse: Elvheim, the hut on the water.

A team of seventy rescuers with searchlights and dogs canvass the mountain and the ridge and the road the night he goes missing and the next day and the morning after that. This will be where we will hope he has walked out, that somehow he came after us and thought, too, that he should first continue to go up. We will hope that he is lost or has broken only a leg, that he is tired, that his camera flash has burned out and that is why he cannot signal to tell us where he is. We will ask someone to drive the road, all the way to the glacier, to see if they see our friend, his backpack or something he has left behind; Luke will run up the mountain, calling for him.

Two yellow helicopters will circle the ridge, back and forth for hours, while the rest of us watch from the stony shore of a glacial lake several hundred feet below. Sam, Ana, Tom, and I will walk up and down the blacktop and gravel road until it turns into highway. We will stop, only for a moment, when Ana says that she feels, in her stomach, *that the way that we make our lives together will not be the same again.* Sam says a prayer, but no one else talks. Instead we watch the beams from the searchers' flashlights travel up and down the dim but still lit mountain for hours, until it is no longer night.

By the next day, a priest has come and two nurses. Friends have caught flights and buses from Bergen, from England, and from France. We talk and talk as we walk along the local roads and sit outside, watching the outline of boats and people and even the shoreline stones reflect off the water. What the psychologist says in the river room, though, is the only thing from those first weeks that I remember. I write his line on a piece of yellow paper, fold it, and put it in my pocket. For many months, I will think about the cost-benefit ratios of relationships. I will recite his line to myself until it begins to sound like an equation.

Although the museum has laid out a guide for what order and even in which direction patrons should walk to view the museum's exhibits, I pay no attention to this and wander at my own leisure and also, out of order, visiting the collections I find most interesting first and later returning to the ones I'm not as excited about.

Somehow, in my personal ordering of things, I leave off one of the exhibits. In fact, I almost don't notice it, but on my way to the gift shop where I'll meet Andrew and Laura I happen upon a doorway to an exhibit I haven't seen. It looks like the room is built for children, and at first I wonder if it's a play place of some sort. As I walk in, I see that it's a room full of experiments, an exhibit titled *Galileo and the Measurement of Time*. A sundial on the ceiling is controlled by a mirror in the window. A cycloid—or arched curve—is traced by a circle pushed along a straight line. There are two models that I find especially interesting. A straight wooden slide is placed next to a curved one, with a ball in each; the sign next to it reads "Brachistochrone descent: a descending body travels faster along the arc of a circle than along the corresponding chord (even though the chord is a shorter path)." Next to this is an inclined plane. I write down a single sentence from its label in my notebook: "This model illustrates Galileo's law of falling bodies."

I touch the models; I try each of them out, placing small metal balls in the top of the different angled pieces of wood and watching as each of those balls moves along the line it's been given, straight for the ground.

I read, too, that day I visit the museum, that the equation $v = 9.81\text{m/s}^2 \times t$ isn't necessarily wrong, but it only works in a vacuum. It doesn't take into account resistance. Air resistance, dependent not only on the friction caused by wind but also on shape and weight, changes the speed of falling; it throws off the standard calculation.

We sit in another room, in someone's white five-story walkup apartment in Strasbourg, France: Tom, Luke, me, the others who had been on the hike. We have come to eat and drink and take a break from seeing the stone house from every south-facing guesthouse window and hearing his name mentioned again and again in shops and on the radio, by people we don't know in supermarkets, in a language we do not understand any longer. We want to sleep too, get past the waking up in the middle of the night sweating through the sheets, feeling the sense that the mountain is there and we are there and he is there, but none of us can ever seem to stop falling, falling.

We drink glasses of wine and spiked cider. There are breads and cheese and curry on plates and candles here too, though perched on the ledge of the window and the white painted coffee table rather than hanging in small jars from the window frames. We've been in France ten days; we are together one last night before some of us return to Norway and some others instead go back home to their old lives in England. We have talked about him a little, though mostly we have talked about our time in France and the ways it has helped or not helped to be in a different place.

There's a break in the conversation, it's quiet, when a friend—the woman in whose apartment we're staying—leans forward and says, *Grief is like a shadow. Sometimes the shadow will be short; sometimes the shadow will be long; you can never predict it.*

Some years after Pisa, a group of students repeated Galileo's experiment. They realized then a surprising truth. It wasn't that the balls were dropping at different speeds but that participants held the iron balls more tightly than the wooden ones, that they took an extra second or two to release.

It is Tom who first thinks he sees him. We are standing in a small dining room, just off the kitchen. I am next to Tom; he is facing the room's only window when he sees someone coming toward the house, across the river on the wooden walking bridge. "It's him!" Tom shouts—the first and only time I've ever heard Tom shout in the several years that I've known him—and sprints outside, in only his green soccer shorts and sock-clad feet, several of the others running behind him.

I watch through the open window, but I realize, several seconds before the guys are across the back lawn, maybe even out the front door, that the man on the bridge is not him. The man's hair is too messy, too long, his frame too compact and thin, his clothes the wrong colors.

Tom makes it almost all the way to the man before he realizes. The moment breaks; I see it swelling: Tom and the clearly wrong man, the mountain behind them, the meltwater river from the glacier crashing against the rocks thirty or forty feet below.

# CAIRNS

One year after the Breheimsenteret—the glacier museum—burned, I returned to that icy valley. I was drawn back to that place in a way that was hard to explain. I thought sometimes of how John Muir described the mountains of California as a sanctuary, a place where there was "an invincible gladness as remote from exultation as fear." The glacier was not remote from exultation or from fear. That's the thing about a world on fire; you wonder if when the match was first struck, what would have happened if you had known to look.

That morning, like so many other mornings on the glacier, was startlingly cold but also bright: a high glare pushing wide swaths of light off the ice, straight onto our bare arms and wind-rapped faces. I could see a reflection in the lake left by the glacier's wake, of the mountain's edge and then more light—sun streaks—moving westward, across the water's thinly ridged surface toward a land where we were less than citizens but more than travelers. I leaned in, against the wind, and tried to gauge the distance between myself and the farthest edge of the glacier; it was a mile, as a bird flies, at least, maybe more, but the ice was still clearly visible, small dunes of white and blue snow rising just beyond the water.

Also in view from the parking lot where we stood, bouncing on the balls of our feet to keep warm, was the hollowed-out space where the glacier museum had once been. Yellow tape was up, which had been blowing in the wind for months, and construction equipment had been hauled in on long trucks to dig out the rubble at the end of the still-standing sidewalk to nothing. A makeshift tent had been set up fifty feet from the site as a temporary information center, with a small café, a few notices about the fire, the timeline for a new building, and some tour guides waiting to help visitors who came by.

We'd stopped inside that tent but only for a moment, long enough to read the signs and to take a photograph of the museum's plan for reconstruction, a photograph that I'd accidentally delete only a few hours later. We'd left for the glacier late that day, and my friends were eager to get onto the ice, to make their way to the other side of the mountain, the side that drifted over the high white ridge and dropped off somewhere beyond our line of sight. I wasn't climbing, only hiking, but I too was ready to get off the bus and start my own walk to the rim of the blue ice.

A small white passenger boat—no more than fifteen feet long—moved steadily toward us across the lake, its metal engine casing bobbing in and out of the water as the captain steered in a half-circle around the rocky bend. The boat stopped just for a moment, a dozen or so meters away, before the captain charged it straight at the shore, close enough I could see the drops of water beaded across its helm. My friends took this as their cue, picked up their backpacks and gear, and scrambled down a narrow break in the bushes and scrub toward the lake and the ever-nearing boat.

They'd decided already to pay for the charter across the lake, an expensive but efficient way to get to the ice. I was in no rush, and so I decided I'd skip the boat this time. We'd meet sometime later, catch an afternoon or evening bus home together.

When my friends and all their gear had been loaded from the shore, I shot a photo for one of the guys and then tossed him his camera. He caught it with both hands just before the boat coasted away, my friends and their guide and the boat's captain becoming smaller and smaller shapes against the brief calm of the morning, against the rush of ice-turned-water.

I picked up my backpack from the ground where I'd set it, adjusted my gloves and hat, and began walking down the single trail that led around the glacial lake to the nearest edge of the ice. Once I was out of the parking area and the lake where I'd left my friends and in the low woods, the wind cut back immediately and I realized, unzipping my jacket, that I'd been straining to hear over it. I wasn't sure what I was listening for—maybe the boat, maybe other hikers—but I heard nothing beyond my own light breaths and the sound of my feet moving along the trail. There were trees and underbrush lining the path. Still, in the gaps between those trees, I could make out leaves, and maybe even small rocks, moving on the periphery of the water.

The word "cairn" comes from the Gaelic *carn* or "heap of stones." As the Greek myth goes, cairns began with the god of travelers, roads, and writing—the guide to the underworld—Hermes. In a debate between Hermes and another god, Hera, over the murder of Argus the all-seeing giant, onlookers were instructed to throw a stone toward the god whom they believed to be telling the truth. Despite the fact that Hera was probably justified in her complaint—Hermes did indeed kill the giant—all the other gods deemed her in the wrong and him in the right. So many rocks were thrown toward Hermes that he was buried under the pile. The first cairn was born: a god inside the stones.

Cairns have long been boundary markers, grave markers, and markers of pathways, summits, and trails. Almost since ships were first built to sail, white-stone cairns were constructed on the edges of seasides, as their own sort of rudimentary lighthouses, directing travelers on their way. European explorers, historian Michael Gaige explains, would even leave messages in cairns, hoping that their friends and fellow countrymen might find them. Future travelers would either destroy the cairns to find the messages (or to reclaim the land for themselves and their country) or would use the cairns for navigational purposes until they found a spot with no cairn where they could set up their own. Of course, cairns come in many forms, and they weren't just constructed to light paths along the sea or to alert and inform future travelers, but also to indicate roads and trails, the tops of mountains, and even to mark the perimeter of long, snow-drifted glaciers, alerting travelers that it may be dangerous to continue on, past the point where they lay.

Paul Basu, another historian of cairns, suggests that in Scottish tradition, the cairn was first a stone of remembrance, that the common Gaelic phrase, "I will put a stone on your cairn," means *I will never forget*. "Cairns embody and invoke," Basu writes, "an effective transformation of a depopulated region into a landscape *haunted* by the memory of loss." In Jewish culture too, small stones were placed on gravesites as a mark of respect and remembrance. Joshua and the Israelites built a cairn before walking around the walls of Jericho seven times, praying and shouting and blowing their horns and then watching the walls fall down and claiming the land. That cairn was a stack of twelve stones, built on the edge of the Red Sea, reminding the Israelites that God had come through for them in parting the waters and would come through again as they moved into the promised land, the land of milk and honey. "To remember," I once hear a Biblical scholar explain—and recall when a student tells me about the Jewish tradition of cairns and about her own experience of placing a small white stone on her grandmother's grave —"is commanded more in the Scriptures even than to obey."

After some time, I begin to make my way to the ice. I'd already hiked the wet and winding three-kilometer path from the parking lot, where my friends caught their boat; I'd followed that to a wooden boardwalk, fixed straight into the side of the mountain, extending outward like a bridge between land and rock. After climbing the boardwalk's many steps, I'd finally arrived in view of the farthest lip of the glacier. A ways after the boardwalk, there was the field of blue ice: twenty- or thirty-foot-high walls of compounded snow, shining and dripping. Beneath the glacial walls was a river, pummeling through the granite and crashing into the lake, the icy glacial water and runoff turning a long plume of water lighter than the rest. Lighter doesn't necessarily mean cleaner though. As I sat midway around the lake on a low sandbar earlier, I had watched the silt slowly wash in across the shore, turning over the water on the small lake like the tide on a sea.

I had taken my time—a few hours, in fact—walking the still-snowy trail, following the bright backpacks of some people ahead and then continuing on my own when I had drifted too far behind the pack.

Even though it's a popular means of getting to the glacier, the trail would have been mazelike save the occasional red *o*'s or arrows spray-painted right onto rocks or trees. The Norwegian Trekking Association uses these red markings interchangeably with evenly stacked stone cairns, indicating paths throughout the country; it's one of the world's most consistent trail systems. Hikes are rated by point values, and small boxes with visitor logs are placed at the ends of trails so that hikers can sign in and record their points.

Hiking is a way of life in Norway. Despite Norway's complicated family-owned private landholdings, it is considered every citizen's right to "roam." The "right to access" or "freedom to roam" law, *allemannsrett* ("all men's right") in Norwegian, means people have rite of passage in any uncultivated area, no matter who it's owned by. It's legal to camp or hike through any property and illegal for landowners to construct prohibitive fences around their property. People can swim in the rivers and forage wild berries from anywhere that they can find them: the sides of highways or someone else's backyard. Land, in this way, is neither wholly public nor private. By law, it lives in a liminal space, somewhere in between. This—and the rise of tourism surrounding the glacier—explains the dozens of people, maybe more, whom I see along the trail, and especially when I reach the river at the glacier's edge. There are six-foot boulders, open ledges, and piles of broken rock, but people walk around them casually, taking photographs and eating bread-and-cheese lunches packed in brown paper bags.

I walk along the rocks myself for some time until I arrive at a single-chain metal fence, thirty feet—maybe more—from the glacier. Metal posts holding up the fence are drilled into the rock. The walls of ice—even in the short distance between—tower overhead. The walls are uneven: webbed in places where the water has melted and pushed through. There are overhangs and bright crystalline tunnels, sunk into the face of the glacier but glinting where the light and shadows glance off the millions of granules of ice.

In the distance, long lines of roped-together hikers make their way up the snow-covered mountain in the not-quite-arctic air. The light moves off those higher places, too, making them hard to look at and hard to look away from, as if watching the midday sun from a dark room. The ice in front of me appears rippled, perhaps the result of shifting, or heat, or maybe just time. Somewhere nearby, there's a sharp cracking sound, and just beyond that is the roar of thousands of gallons of meltwater moving, creating channels below surface.

I walk closer. I look to make sure no one is watching and then jog to the top of the rocky ridge and then down, under the chain fence, past the *Caution: Beware of Falling* sign, and, finally, straight to the walls of ice.

Perhaps the most famous of the cairn exploration stories was the search for Sir John Franklin's lost cairns on Baffin Island. In 1845 fifty-nine-year-old English Royal Navy officer John Franklin and 128 of his men sailed from Greenhithe, England, to Baffin Island on two ships, the HMS *Erebus* and the HMS *Terror*, hoping to traverse the last untraveled section of the Northwest Passage to find a westward route from the Atlantic to the Pacific Ocean. By the time Franklin's crew sailed, the Northwest Passage was a route that had been searched for unsuccessfully for almost three hundred years, including attempts by Martin Frobisher, Henry Hudson, and James Cook.

A mere six years after Franklin's expedition began, a first route across the Northwest Passage would be discovered, but this wouldn't be by John Franklin or by his officers or his men. Despite being a celebrated and experienced explorer, Franklin never returned to England. His ship was last sighted by another only two months after it had left Greenhithe and before it had made it into the tip of the Northwest Passage.

By 1848 search-and-rescue parties were sent to Baffin Island to search for cairns or any other marks of the lost ship. The search continued for some years, with as many as thirteen boats at once looking for traces of Franklin's ship or crew near Baffin Island and the parts of the Northwest Passage that followed. In 1854 one explorer, John Rae, brought back stories from Inuit about the ship's dying party, including the tale that Franklin's men had, in the end, resorted to cannibalism. People in England didn't believe Rae or his sources and the idea that Englishmen would resort to cannibalism, though, and Rae's name was blacklisted, some of his own maps even then credited to another explorer.

It was in 1859 that the first reliable trace of Franklin's voyage was found. That year another explorer, Royal Navy lieutenant William Hobson, discovered several cairns on King William Island, built by Franklin's shipmates. Inside one of the cairns was buried a tin can with a paper message:

28th of May, 1847

H.M. *Erebus & Terror* wintered in the ice in Lat. 70° 05′ N, Long. 98° 23′ W

Having wintered in 1846–1847 at Beechey Island in Lat 74° 43′ N. Long. 91° 39′ 15″ W. After having ascended Wellington Channel to Lat. 77°, and returned by the west side of Cornwallis Island

Sir John Franklin commanding the expedition.

All well.

As David Williams notes in his book *Messengers in Stone*, all was not well for long. The note in the cairn was updated to add that eleven months later the ships were stuck in ice, and already Sir John Franklin and several other crew members had died:

> April 25, 1848—H.M. ships 'Terror' and 'Erebus' were deserted on the 22nd April, 5 leagues N.N.W. of this, having been beset since 12th September, 1846. The officers and crews, consisting of 105 souls, under the command of Captain F. R. M. Crozier, landed here in lat. 69° 37′ 42″ N., long. 98° 41′ W. Sir John Franklin died on the 11th June, 1847; and the total loss by deaths in the expedition has been to this date 9 officers and 15 men.

In recent years there has been a move to stop people from adding cairns to wild areas or to altering the cairns that are already there. Park rangers and naturalists are concerned that the building of additional cairns can disrupt the landscape, causing soil erosion, changing the habitats that many insects, animals, and plants live among, and leaving a definite trace of human intervention in the wild, marring the experience of wilderness hikes and climbs for other travelers who will one day come.

There's also a concern that unauthorized cairns might misdirect hikers and backpackers, especially in bad weather. In "A Natural and Social History of Cairn Building and Maintenance," Michael Gaige offers the study of an Arcadia National Park ranger, Charles Jacobi, as evidence for the trend of changing cairns. Between 2002 and 2003 Jacobi studied the cairns in Arcadia, noting their movement and modification day by day. Even with informational signs asking hikers not to build or move cairns, Jacobi found that on average, 23 to 39 percent of cairns on popular hiking trails are altered every five days. "The cairns on top of Dorr and Sargent mountains are twenty feet wide and eight to ten feet high," he said in an interview. "It would fill my office. And all those rocks came from somewhere on the mountain. All of them at one time served as a habitat for plants to grow around and spiders and other invertebrates to get cover.... I think most people don't think about that, but I do."

Other hikers and naturalists disagree, seeing cairn-building as a profound and participatory act of meaning-making, that to outlaw the practice of building cairns means to lose something of the helpful, communal nature of exploration, to regulate—too much—the wild. In the last several years cairn building has even become its own form of temporary art. British land artist Chris Drury constructs cairns of carefully sculpted and placed rocks in a variety of landscapes. In one installation, *Covered Cairn* in Langeland, Denmark, Drury built a dome of woven sticks on top of a stack of glacial rocks; in photographs, snow falls on the edges of the dome, making the cairn look like it's ensconced in fog. Andy Goldworthy, too, has recently set up installations of cairns in England and Australia. His rock statue *Strangler Cairn* famously cost the Australian government nearly $700,000. Goldsworthy used hand-cut pieces of granite and slate in the ten-foot installation. He planted a fig tree inside that would eventually overtake the stones around it, an attempt to portray both the beauty and the violence of nature.

Of course, it's not only stones scoring landscapes either. There are orange blazes throughout many trails in North America, black or red *x*'s in much of Europe. In Germany, some trails are given their own logos: witches' hats or wine bottles or bolded trail names fixed or sprayed onto important directional points. Before the 1785 Land Ordinance, people in the United States delineated land boundaries and even paths by notching with an axe trees along the perimeter. They would walk back to these trees every year to reaffirm their grasp on the places where they lived and worked. As Kent C. Ryden notes, in *Mapping the Invisible Landscape*, these premapped borders were called "witness trees."

I walk along the glacial ice for some distance before realizing that I haven't—in any way—recorded the spot where I'd climbed down. Sometimes I leave two sticks in an "X" on some early point on the trail or take a photograph that I can look back on of some significant navigational detail. It's a basic tenet of responsible hiking: mark any turns; take note of important landmarks; don't forget where you've come from.

This day, drifting along in my own thoughts in the unsteady landscape, I've done none of those things. Though I can still hear the river and see climbers, much higher up, making their way along the crest of the mountain, near the very tip of the glacier's ice forms, I can't remember what the ice looked like in the place where I first studied it or even where I left the *Caution: Beware of Falling* sign. The tunnels of blue and long expanses of densely packed snow have already begun to run together in my memory. I remember noticing as I walked that each view I saw was different than the last, but not with the sort of clarity that marks a difference between their determinate edges. I realize too that I can't remember exactly when I last saw someone walking or hiking nearby. There was a family taking photographs when I first walked down to the walls of ice, and I'd seen some young guys clambering down the same initial path I had. We'd diverged at some point, though I hadn't noticed when or at what spot.

I consider continuing anyway, to see what's just up ahead, if there's a better view of the valley I came from, or possibly a clearer path back to the parking lot where the trail started. My friends and I haven't set a meeting time, though, and I don't want to delay or worry them, so I turn back and begin walking roughly in the direction in which I think I've come.

After a few minutes of walking, I decide to try scaling a small ledge. It's gray and beveled and oddly shaped, almost like a horseshoe, the top edge of the ledge smooth and jutting out over air but the bottom cracked and pitted. I grab the top of the ledge with both hands and, in one quick motion, hoist myself onto it and then stand up. From the top of that rock I can see several other ledges, like stepping stones, but all six or eight feet above my head. There doesn't appear to be any path I can reasonably get to, without straight vertical climbing. I jump back down and walk farther in the direction I've come, but I can't find anything else to climb. Mostly, the uneven walls of rock are too high or too loose to safely scale.

Finally, I notice a pile of scree—broken stones—that seems to jog my memory. The stones are of all different sizes and angles, much like what you see at the very top of most mountains. I start walking slowly up the pile, but the stones aren't steady, and so as much ground as I've gained, I almost immediately lose, slipping right back down. I decide to move a bit faster; I stride and scramble up the pile, causing small avalanches of tiny stone and dust with the weight of my feet, but it works; I make progress. After a minute or two, I get to the top of the pile of stones and find level ground—a broad, flat expanse of rock—just beyond it.

The sky is a bright, almost crystal, blue, and I can see the glacier even more clearly in the distance. There are no people though and no fences, just melting snow and a theater of rock, lining the high ledge.

Sir John Franklin wasn't the first to sail the HMS *Erebus* and the HMS *Terror*, looking to traverse unexplored lands. These same two ships were used only four years earlier by Sir James Clark Ross in what's now known as the Ross expedition to Antarctica.

In October 1839 the ships set out from England to Tasmania and then to Antarctica, looking to discover the South Magnetic Pole. On January 1, 1841, Ross and his men arrived in the Antarctic Circle. Sailing through open seas and ice, the expedition discovered the Ross Sea, the Ross Ice Shelf, Victoria Land, and two volcanoes, Mount Erebus and Mount Terror.

In his personal narrative of the expedition, Robert McCormick, the expedition's doctor, naturalist, and geologist, would write of discovering Mount Erebus:

> Thursday, January 28, 1841.—We were startled by the most unexpected discovery, in this vast region of glaciation, of a stupendous volcanic mountain in a high state of activity. At ten a.m., upon going on deck, my attention was arrested by what appeared at the moment to be a fine snowdrift, driving from the summit of a lofty, crater-shaped peak, rising from the centre of an island (apparently) on the starboard-bow.
>
> As we made a nearer approach, however, this apparent snowdrift resolved itself into a dense column of black smoke, intermingled with flashes of red flame-emerging from a magnificent volcanic vent,

so near the South Pole, and in the very centre of a mighty mountain range encased in eternal ice and snow.

In the end, the Ross expedition didn't make it to the magnetic pole. Ross himself, though, did make it to the North Magnetic Pole, where he constructed a cairn and added to it a British flag. He was also the first to set out on an expedition to search for the lost John Franklin and his own former ships.

Between Krossbu and Turtagro—two of the highest elevation areas near the Jostedalsbreen Glacier—there is a ten-kilometer stretch of three hundred cairns, each with a metal pole set into the rocks to help travelers go from one point to the next on the high, cold plateau. Until the famed 2009 Full City oil spill in another city in Norway, Molen—the Norwegian word for cairn—more than two hundred cairns of varying sizes rested on the local shoreline. These cairns, my Norwegian friends tell me, have been there for centuries. They once indicated graves for warriors, lone roads across the high mountains, or the clearest path along the shore of the sea.

I didn't get to see the cairns at Molen during my time in Norway, but I did make it to the highway of cairns between Krossbu and Turtagro. I was with friends, Tim and Annie, and their two children that day; they were visiting from Manchester and had already driven twice that afternoon into the snowy mountains but offered to go one more time and to take me along.

We layered in wool socks and sweaters and the closest things we could find to winter jackets, and then Tim drove us up the mountains, past the tree line and the browning grass and the herds of wandering sheep, to the place where the cairns began, one stone stack after another. We could see the high mountains up there, almost from every angle, our car a satellite rotating between them. At one point, the snow in one of these passes was several feet taller than the car. A single path had been sliced through it, just wide enough for a road.

When we got to the drive's highest point, Tim parked the car on the side of the road and we all got out. There were dips and rises in the land around us and mountains off in the distance. A cold wind cut across the open space. Though it was summer, the ground was mostly frozen, with a few patches of mud and a thin layer of snow covering the rocky outcrops and the high, treeless plain. There was a dark, narrow lake not far from where we had stopped, but otherwise the landscape was cast in shades of brown, gray, and white.

While the kids made snowballs to throw at each other, I walked up the spine of a small hill. It was only fifteen or twenty feet higher than where the others were, but I could feel the air through my jacket on top of it and could hear it moving too, though there was nothing in that wide, barren landscape to rustle. Low, thick clouds hung over the tops and sides of the mountains, and a fog was coming in from the east. There were no people other than us, no cars, no cabins, no houses or hotels or churches. In the shadow of the mountains, on that high alpine plain, there was us and there were small piles of rocks, of varying sizes and heights, all across the land.

I'll read somewhere later that these particular cairns—which follow an east–west route all the way across the mountains—were set up as "guideposts" for farmers and tradesmen crossing the dangerous pass. It wasn't until the nineteenth century that the cairns began to be visited as an attraction rather than used as utility, and it wasn't until 1938 that two hundred young men used "pickaxe, spade, crowbar and wheelbarrow" to construct the highest mountain road in northern Europe, in the place where there were only stacks of stones before.

I walk along the flat ledge, unsure where exactly I'm going but realizing that if I move in the right direction, this makeshift path will probably lead down, maybe to the wooden boardwalk or the muddy trail, to the long glacial-fed lake, the parking lot, to the highway and the handful of cars that travel on it, or to the low green trees just ahead of where I stand.

I've hiked often enough and taken enough wrong turns not to worry too much, not as long as I'm in a clearing or somewhere with a fair bit of open air. In the woods, losing one's way feels more perilous; even if villages or towns or cities aren't far, there's a sense sometimes of the trees closing in on you; the same place that minutes before felt like a solace can suddenly seem like it's engulfing, muffling the light and air and sounds of whatever you're looking for, or whoever might be looking for you. It's different in the open air; there's so much more possibility of seeing and being seen, even if you're in the middle of somewhere that's nearly silent.

It's warmed up some, and so I take off my jacket and stuff it in my backpack; I take a drink of water from the bottle I'd filled up earlier while I was sitting at the edge of the river that joins the glacier and the lake. Though it's been an hour or more since then, the water is still cold. After a long drink, I put the bottle back into my backpack too and walk on, concentrating, this time, on paying attention to what's around me: the shades of gray of specific rocks, the angle of the mountains behind and beyond me, the shapes of blue ice in the distance.

Occasionally, the ledge seems to split, the large rock breaking into smaller pieces or joining with other massive granite shelves. I keep walking until soon one of these shelves begins to gradually lead upward. It's just a little higher than where I'd been before, but this time the ledge breaks into a wide clearing. It slopes slightly, but the view is suddenly open, straight to the steep bank of the mountain ahead. There's sun too, sweeping across the expanse of rock.

When I was a child, my father used to read me a story about cairns. They weren't called cairns, but I realize now that's what they were. A young boy and girl in that story were tasked with finding a lost prince, who was buried underground under a witch's enchantment. To save the prince—and the world itself from the witch's evil schemes—the children must remember four landmarks, four signs that they've been given: watching for an old friend who can help them along their journey, finding a ruined city of giants, seeing a message written in stone and following its instructions, and knowing the lost prince by a phrase that he says.

The children, at the end of the story, do discover the lost prince but only after significant lapses of memory where their own desires, sharp smells of food, and bright invitations by the cunning witch and even trying-to-be-helpful strangers make the lights of their memory seem dim and hazy. The signs are not quite what the children thought they might look like, and it becomes tiring, at points, to continually recite them.

The story was fitting for my father and I to read, both of us having the worst memories in our family, forgetting names and places, sometimes the events of whole weeks or summer vacations. For some reason, the story of the boy and the girl mattered in a way that the other stories did not. I thought often about that boy and girl and what had been required of them in order to find the prince and save the kingdom. Long after the book was over, late at night, lying awake in my bed, I'd ask myself, *If I were one of those children, would I have seen the signs?*

I wonder now if John Franklin and his men saw the signs. Their ships, after all, were named the HMS *Terror* and the HMS *Erebus*, *Terror* being a throwback to when that ship was a warship used in the War of 1812, and *Erebus* named after the Greek god for darkness, the son of chaos, and the entrance to the underworld.

John Franklin's wife was famed to have demanded a search for him, eventually helping launch seven expeditions to look for him, his missing men, boats, and any possible records that he might have left, buried in cairns. She also contacted celebrities of the day, including Charles Dickens, to try to advance her cause, traveled herself to the north of Scotland—the closest she could get, at the time, to her husband—and even went to a seer to try to get news of Franklin.

In the end, it wasn't until 2014 that the HMS *Erebus* was discovered, near King William Island in northern Canada, by a boat sent from the Canadian government. Two years later the HMS *Terror* was found by the Arctic Research Foundation, incidentally, in Nunavut, Canada's Terror Bay.

I'd talk sometimes in those days in Norway about life back in the States as "real life."

"This is your life," my friend Annette leaned over and said once, "as real as it will get."

It was strange to have found in Norway a place that felt so much like it was my home, though it was not the place where I'd grown up or would return to at the end of all those summers and years. I was walking in the places where my great-grandfather had once walked and the places my grandfather still dreamed about, but they were not the places of my parents and they were not my own. When I first went to Norway, I hoped for but did not expect a home or even an embrace; I thought I was coming with open hands to the place, but I did not realize that sometimes opening one's hands to one thing means closing them to another.

Even the cairns are perhaps whiter in my memory than they were at the time: the white of birds' bellies, the white of sky, the white of bleached wood after rain, the white of frozen rivers, in pale early-morning light.

It's in that clearing where I first see one, only slightly higher than the earlier vantage point of my eye: a pile of small rocks, not far off in the distance. The rocks in that pile are misshapen and barely balanced but smooth, no doubt from years of wind and rain. One stone is wedged on top of another, tightly, each a different hue of granite, a slightly different patch of mountain. Stacked there in the middle of nothing but rock and sky is a cairn, stone on stone. I reached down and touched the top stone on that first cairn I saw on the glacier, but I did not move it, did not take any stones away or add anything to the stack. It was knee-high; the stones were flat and mostly smooth, the cairn broad at the base and narrower and thinner as the small flue rose upward, the sun behind it flaring into the rocky valley below, light: light that could melt the snow, crack the ice, dissolve the rock into a wide, open plain.

After that first cairn was another. It was a bit smaller, tight and low to the ground, with nearly whitewashed rocks that were layered in a circular formation. The base stones were mostly large—perhaps the size of bricks—with the upper stones increasingly smaller like on the first cairn, but this time pebbles were wedged into the open spaces of the cairn, shoring it up against water or wind or people walking by. The outer perimeter of the cairn was arranged in almost perfect symmetry, the edge of one rock exactly meeting the edge of another. Beyond its stones was a view of the valley: scraggly bushes, mosses, wild grass, and, in the distance, dense groves of trees.

I walked on, following as best as I could the angle of the cairns down the steep and curving slope of rock. Just when I thought I'd missed it—when I contemplated hiking back up to the last pile of cairns and starting again—there was another cairn stacked low to the ground, and another. Standing in a rough line, like evergreens, were several others. Each cairn was a similar height and composition, each within eyeshot of the next, each someone else's—or several people's—essays of rock, attempts to map the way forward or behind, perhaps signaling the way home or the farthest reach of where the glacier once went before the warmer years, before the retreat of the ice. There were millions of cracked stones and then, rising from them, cairns.

The cairns—and the place—marked the way, yes, but what way they marked I did not know. I followed them anyway, feeling a paradoxical weight of gratitude and longing. One stone spire at a time, I moved over rock piles and down ledges, past walls of ice, scrambling sometimes, moving quickly always from one stone to the next. Each of the piles of cairns led to another within view, and each gradually led down, off the glacier, and toward the low lake where I'd begun.

At the end—arriving at the last pile of cairns, a small stack of just a few white stones, square mostly and low to the ground—I found myself on a wet, grassy knoll, looking back out over the lake and the dark trees and the silt and sand. I was not where I started, but from where I'd come I could see a trail and it wasn't far off.

# TO THE CENTER

I

"It was Sunday, the 24th of May, 1863, that my uncle, Professor Lidenbrock, came rushing suddenly back to his little house in the old part of Hamburg, No. 19, Königstrasse. Our good Martha could not but think we was very much behind-hand with the dinner for the pot was scarcely beginning to simmer." And so begins the Frenchman Jules Verne's extraordinary voyage, *Journey to the Center of the Earth*. The first English version of the story took some liberties in the translation. That version starts, "Looking back at all that has occurred to me since that eventful day, I am scarcely able to believe in the reality of my adventures. They were truly so wonderful that even now I am bewildered when I think of them." The persona of the narrator changes with this shift from a simple relayer of information to interpreter of it. A note of disbelief is added, and an identification with the reader. It's the word "truly" my eyes rest on, though, when I read the English version for the first time. "Truly" *implies room for lack of truth*, I remember a professor once telling me, *suggests that facts here can't speak for themselves.*

The facts of the story are that Verne's protagonists, Axel—a young student—and his uncle, Otto Lidenbrock—a professor, scientist, and savant—uncover a hidden runic manuscript that falls out of a book. Once decoded, the runic manuscript takes them from Germany to Copenhagen and then across the sea to the snowy wilderness of Iceland, glaciers all around. Finally, they journey to an extinct volcano that the writer of the rune, Arne Saknussemm, assures them will lead to the very center of the earth. "Descend the crater of the Jokul of Snäfell," the ancient rune reads, "that the shadow of Scartaris softly touches before the Kalends of July, bold traveler, and thou wilt reach the center of the earth, which I have done." Axel and Lidenbrock and their trusted Icelandic guide, Hans, do descend the crater and, after a series of adventures and misadventures, find themselves at the center of the earth.

II

I first listen to *Journey to the Center of the Earth* on a ten-hour drive to Wyoming. I'm planning to teach an intermediate British literature course themed around science and empire a few months later and have decided we'll read Verne alongside *Gulliver's Travels*, *Robinson Crusoe*, and *The Voyage of the Beagle*. I download the entire book—unabridged—and burn it onto a set of six or seven silver CDs. As my car crosses the low highway through Kansas farm fields, beside combines and four- or five-foot-high stalks of corn, I listen to Axel narrate his trip from Germany to Copenhagen to Iceland, from making it to the edge of the volcano to descending with the help of ropes, strange acoustic phenomena, good conjectures, and the ever-faithful Hans into its heart.

"Is the Master out of his mind?" the housekeeper asks Axel while Professor Lidenbrock frantically scurries about, making preparations for their journey.

Axel nods.

"And he's taking you with him?" she asks.

He nods again.

"Where?" she asks.

"I pointed toward the center of the earth," Axel recalls.

"Into the cellar?" exclaimed the old servant.

"No," Axel said, "farther still."

After some time, the tractors and silos and eventually even fields outside my window subside. They're replaced by arid ground and boarded-up sheds. Towns and villages and filling stations come less and less often until they almost disappear altogether. When I open the window, heat surges in, weighs on my leather steering wheel and the backs of my hands. My car moves steadily still, the engine pulsating warm, the road outside a blur of similar lines against the moving interior.

And then, hours later, there are mountains, in the distance at first, and then closer and closer still. I pull over on the side of the road when I can first see them. I shut off my car and walk down a dirt road beside the highway for a little ways. A pickup truck flies by; a dog barks somewhere barely within earshot. Otherwise, it's me and the dusty stones and an empty field and the mountains in the distance, hanging over the edge of the horizon like a precipice.

### III

I walked and walked the only summer I lived in Wyoming. I was staying in a professor's house in town: tan stucco with terracotta tiles and a built-in wooden table in the kitchen and skylights in the upstairs bedroom, but no working lock on the front door. There was a bicycle out in the garage, the professor had told me before she left town for the summer, and I put most of my own things in storage and moved in. I can't remember now whether I ever even opened the garage door.

I walked to friends' places; I walked downtown; I walked to coffee shops and restaurants and to the doctor's and the local park and the laundromat. And when I didn't have someplace to go, I walked around the tree-lined neighborhoods by the university, trying different blocks each day, sometimes carrying a folded guide to Laramie in my front pocket, tracing on it in pencil the places where I'd been. I walked around the wide oval-shaped pasture at the center of campus and walked every route I could to my nearly empty office with its own long wooden table, taking different footpaths there, even different staircases inside.

I walked, nearly every day, to a vegetarian café downtown, even though I was not a vegetarian. I would sit in a creased red booth in that café and order the same lunch, sometimes with a book or a pen and paper, sometimes just watching the people move along that broad, open street, a few glass-windowed stores and restaurants on one side, nine or ten sets of railroad tracks on the other. Then when I was done, I'd walk twelve blocks across town, past the post office and the edge of the university campus, to a yellow-painted coffee shop. Sometimes I'd just walk by; sometimes I'd go inside and buy a drink.

I walked in the mountains and at state parks too. I walked along the red rocks of a high climbing park, Vedauwoo, and noticed, for the first time, all the pines browned or whitened, eaten from the inside by waves of beetles. I saw a family of bears once on a steep walk in the mountains; another time, I turned at a fork in the trail and there, maybe three or four feet away, was an adult moose.

## IV

One of the legends about Verne's life—probably not true in whole but maybe in part—was that when he was a young boy, ten or eleven, he is said to have sought out a spot as a cabin helper on a ship planning to sail to the Indies. He wanted to retrieve a coral necklace, his biographer niece explained, for his cousin Caroline. As the legend goes, Verne made it onto the boat but was caught by his father before the ship set off for the next port.

In another story the young Verne rowed his own small boat from the local harbor, trying to catch up with a larger ship sailing out to sea, but when caught was made to promise that, while still a boy, he'd only travel the world in his imagination.

Imagination he had. Verne had heard stories of adventurers from his teachers; whatever the cost, he wanted to find himself among them.

As an adult, Verne would spend days in the National Library, carried far off by research on geography and science and reading others' travel narratives. He'd meet with famous geographer Jacques Arago, sail around Europe himself, buy boats and name them after his son, the *Saint-Michel I*, the *Saint-Michel II*, the *Saint-Michel III*, and say that he wanted to invent the novel of science.

## V

My students and I talk about context, about thinking through our own and others' lenses as we read. We talk about Lyell and Hutton; we read about the Victorian turn from a religious young earth to an older geologic record. We talk about imperialism, colonialism, power, the nineteenth-century deists, and the rise of science in the cultural discourse.

My students and I try to decipher, to decode, why so many years after the fact, so many people still read *Journey to the Center of the Earth*. One student makes an argument that it's Verne's reliance on classical Roman and Greek mythology. "It's everywhere," she tells our class, pointing out each Greek and Roman reference in the text that she's found, the rest of us flipping along as she calls out one page number after another, ending with her strongest evidence that this is why the travelers eventually land in Stromboli, Italy. Another student comes to class with a huge map he's hand drawn: the earth according to Verne's novel. Certain sections are colored in with colored pencils or carefully labeled. At the middle of everything, in the middle of that student's drawing, the earth is hollow.

We talk about how Freud might read this text and Said and Marx and how it contributes to, works with, and works against the literature of exploration. My students talk about the descent into the self, the descent into logic, the descent into science, the descent into madness, about madness, about machines, about the fossil record, about the emotional registers of the savant and the scientist and the student, and the way the story doesn't end where it begins, that Verne sees to it that the protagonists go somewhere.

One student throws a viewing party for a movie adaptation of *Journey to the Center of the Earth*. Someone rents the DVD, someone else brings popcorn, and my students come back the next Monday telling me how, strangely, Axel was not Lidenbrock's nephew in this version, how there are villains added to the story, and also how Axel and all the boys at his university seem randomly to break out into song.

## VI

That summer in Wyoming, sometimes my friends Julie and Paula and I went walking together in the Snowy Mountain range, an hour or so past town. One time, my car barely got up to the mountain parking lot, the accelerator jumping when the car went downhill, stalling as it went up. But we did make it there, to a trail in a quiet evergreen forest. We took a photograph before we began, balancing someone's camera on the roof of my car, and then running to get into the shot. Paula's on the left; Julie's on the right; I'm in the middle. They're both taller than me and dressed better for a hike, Julie with a backpack along and Paula carrying food and water in a bag on her hip. I have a water bottle in my hand but nothing else beyond a green cotton shirt and a single car key in the pocket of shorts which that summer had become much too big.

"Be careful," my doctor had warned me earlier that week. "You're getting too thin." Amy, a slim thirtysomething with plastic-framed glasses and fine blond hair pulled back loosely, was sitting on a stool while she said this. I sat on her table, the same way I had nearly every week for the past four months, next to the sign that read "Health Care is a Right, not a privilege." We had both thought after a host of inconclusive tests that maybe it was stress, though sometime later I would discover it was likely a deer tick, burrowing down into my body.

I was careful, but on this day I was walking.

In the photograph we're small and distant, set against a line of trees, tall grass, and blue sky.

The trail crossed a bridge over a small, snow-fed mountain stream almost immediately, tiny shoots of whitewater pummeling rocks and moving fast, downward. The path beyond the stream was dry and narrow, millions of needles layering the ground and trees all around, rising upward, beyond our view. It was bright still—even with all those trees—spots of sun regularly making it through breaks in the branches and onto the trail or our own moving bodies.

We talked as we walked around bends and up, over fallen trees. We continued past one long stretch of loamy mud, trampled in by large animals sometime before. We walked alongside uneven boulders cut into the trail and hundreds of aspen with their paper-white trunks, scattered Douglas fir, and thousands of lodgepole pine, some green with branches stretched out, others dead or dying.

We walked up and up. There had been no one else in the parking lot and there was no one else on the trail, only us, moving deeper into the woods, farther into the mountains, no map in hand.

## VII

The only thing the professor asked of me was that I keep her grass alive. The mail had been forwarded, someone hired to mow the lawn, the answering machine turned off, her friends and colleagues notified of her travels.

It seemed like an easy task. When she left and I moved my things into the upstairs bedroom—a suitcase, a laundry basket, and a box of books—the grass was full and green, the sort of green that's bright, unnatural almost in a high-altitude plain where the sun beats heavy and the wind comes through long and slow. The grass in her yard was soft underfoot when I moved in; I got into a habit of sitting outside in a patio chair, my feet resting in the grass, in the early mornings when I'd eat breakfast. I'd scramble through it sometimes too, on my way to water the flowers in the afternoons. She had told me she had no bones about whether the rest of the plants stayed alive, "just the grass," she'd said when she showed me the house a few weeks before the summer. I watered the flowers anyway.

We walked, that day she showed me the house, from the university to her place so I knew where I was going. We walked from the small entryway—coats and hats hanging on racks—to the living room with its large fireplace and piano, through the reading room, kids' artwork all over the walls, into the kitchen, downstairs to the basement washer and dryer, upstairs to the den with its worn sofa and chaise and basketball hoop inside, to the black-tiled bathroom with national park posters hanging next to the tub, out again to the large white bedroom with its single round window facing the front yard.

It was different—that house—than my own single-story, open-plan apartment. Those first weeks in Wyoming I was always walking room to room, in and out, up and down. It felt like the house was located in the very center of everything. I could walk anywhere from there: Julie's house, Paula's apartment, the local Episcopalian church where I'd sometimes attend services, the park where I'd often sit and read.

## VIII

Our class meets in a basement room with no windows, so even when it's not movie night it feels subterranean, like we're all always descending to get there, but descend we do, into the text, into the ideas, into the room. Sometimes we descend so far, I wonder if we'll ever come up for air. We talk about many scenes from the book in great detail, but the scenes my students are most interested in are the dream scenes, the scenes where the narrator of the story makes a descent into night, where he becomes untethered in some way from his own reality.

I would guess that my students don't know what it's like to cross an icy country on foot, to descend a volcano, or to ascend another, but they do understand what it's like to wake up in a cold sweat, trapped in their own minds, in their own lives or apartments. I'd guess they also understand too a little of what it's like to see something for the first time, to realize the cellar is not the only thing below.

"If at every instant we may perish," I quote to them from Verne one day in that basement classroom, hoping it consoles them, hoping it's comfort, "so at every instant we may be saved." I try too to secure the class another classroom, "something aboveground," I mention in my email to the room scheduler, "or one with windows." The person who replies tells me that the room across the hall—the same configuration as the room we're in—might be open but that everything else on that side of campus is taken.

We stay. We walk down together sometimes, a few of us at a time, following the same path from my office, past Speaker's Circle and the back of the business school, past some wooden benches and the side of the building where students often smoked before and even after cigarettes were banned, and down, into our basement classroom.

## IX

"The whole fossil world lives again in my imagination. I go back in fancy to the biblical epoch of creation, long before the advent of man, when the imperfect earth was not fitted to sustain him. Then still farther back, to the time when no life existed. The mammifers disappeared, then the birds, then the reptiles of the secondary epoch, and then the fishes, crustaceans, molluscae, articulata. The zoophytes of the transition period returned to oblivion. All life was concentrated in me, my heart alone beat in a depopulated world. Seasons were no more; climates were unknown; the heat of the earth increased until it neutralized that of our radiant star....

"Ages seemed to pass like days! I followed step by step the transformation of the earth. Plants disappeared; granite rocks lost their hardness; the fluid replaced the solid under the influence of growing heat; water flowed over the earth's surface; it boiled; it volatized; gradually the globe became a gaseous mass, white hot, as large and luminous as the sun.

"In the center of the nebulous mass, 14,000 times larger than the earth it was one day to form, I felt myself carried into planetary space. My body became ethereal in its turn mingled like an imponderable atom with the vast body of vapor which described its flaming orbit in infinite space!

"What a dream! Where is it carrying me?"

## X

In the dream I'm falling. In the dream I'm lying on the attic floor and then I'm not. In the dream it's a clear day, always, the sky a systematic shade of blue: blue like the reflection off a salt pond, blue like blood, blue like a wave impending, blue like the irises in my own eyes, in my father's eyes, in my mother's eyes, in my brother's eyes, in the eyes of only one of my two nieces.

In the dream I'm fine, always, and then, in the dream, I'm not. In the dream I never see the lip, the cliff's rim, the mountain's drop, the perfectly perpendicular edge of the top of the glass-sided sixteen-story building, only air, distance dissolving.

In the dream I'm neither floating nor being carried along, just moving. In the dream it's neither cool nor hot; in the dream I don't see the ground; I don't see the leaves on trees or people walking along the sidewalks or street signs or rivers or valleys or anything below. I only see sky, the sky ahead of me, all around me—but never touching me, never penetrating my skin—fading from a bright blue to a pale, unnatural gray. I never touch down in the dream; I just fall and fall until I jerk awake and realize I've thrown the white cotton sheets off the bed again.

I watch the moon through the skylight or sometimes the dim morning light flinging shadows around the bedroom floor. After a while I get up; I drink a glass of water. I open the window and suck the cold air into my lungs. I plant both of my feet onto the ground.

In the dream, I realize early one morning, I'm never walking.

## XI

Axel and Lidenbrock walk and walk in *Journey to the Center of the Earth*. At first, when the pair arrive in Iceland, meet their guide Hans, and begin their quest to reach the center of the earth, they ride on horses through the country, lodging with families in one small village and then another. After a while, though, the ground becomes too difficult for the horses to smoothly traverse and the travelers continue on foot. Axel and Lidenbrock and Hans walk in single file by fjords and over a bog; they walk over long stretches of basaltic rock and past massive blocks of stone. They walk all the way to the base of a mountain, and then they walk to the top of that mountain, five thousand feet above the sea. Once they reach the summit they locate the ancient volcanic crater Arne Saknussemm's rune has led them to find. They descend that crater, make their way to the very bottom of it, and then they keep walking, into the black, down toward the center of the earth.

Before Axel and Lidenbrock even get to Iceland, they practice walking. They walk through the narrow streets of Copenhagen and up to the top of the tallest steeple in the city, the Von-Frelsens Kirk. They circle up and up the spiral staircase, around the steeple's spire. They walk inside first and then, after 150 steps, continue outside, on the very edge. Axel is giddy as they walk; he is sick; he exclaims, "I shall never do it!" He climbs on his legs and then his knees and then his stomach. He sees the city below covered in thick smoke, sees the universe spinning.

Then finally, when he's made it to the steeple's farthest point, he descends each of those steps, back to the ground. And then he tries it again, five times more in the next five days. "My first lesson in vertigo lasted an hour," narrates Axel, "and when at last I was allowed to descend and my feet touched the solid pavement of the street, I was lame."

## XII

Gradually, that summer in Wyoming, I began to walk farther and for longer, eventually beyond the city's bounds. I moved from one coordinate to the next; I mapped the landscape with my body. By the end of the summer, I began to go nearly every day to a high alpine park, Happy Jack Park, where I'd walk the Ridge Trail, the Meadow Trail, or the Campground Loop. I wasn't sure what I'd find in that place, but I hiked it still, wanting to figure it out as I went along. Sometimes there were animals—cows, birds, deer— or other runners or hikers, but often it was only me, walking quietly and still slowly, through the low brush and trees and high dirt paths.

On the best days in Wyoming, I'd walk Happy Jack with my favorite professor, Kate, and her bright black dog, Clara, or with Julie and Paula. We'd walk through a high alpine forest, tucked just far enough away from the main highway that people who were not locals did not go there. It was a dark woods, the red of hemlocks and cones and needles from pines and spruce padding the air and the ground. There were a few hikers and animals on those days too, but mostly it was quiet, the white sky occasionally making it through breaks in the trees, stripping the shade bare. It was cool; there were forget-me-nots, wild violets, a few small campsites and bike tracks, but otherwise narrow footpaths and high-altitude air, away from the pressure of the city and the buildings and the relentless streets. We'd walk and talk, and if Clara was there she'd lope ahead, a flash of black dodging in and out of trees.

One day I went on the Summit Trail at Happy Jack and tried to make a big loop but accidentally ended up on another trail at the wrong parking area. I turned back the way I'd come and wandered one trail, then another, until somehow I ended up off the trail entirely, looking for my way back as dusk quickly closed in on me.

It had been cold when I started hiking, rain threatening in the low clouds on the horizon but not yet touching down. The trail had been wet in places, wide patches of dark mud, packed down by livestock and bicycles and other hikers and making the route almost impassable. The woods were even darker than usual that day, and the wind seeped through the loose weave of my jacket. I continued on anyway, shaking off the cold. I saw only one person as I went along. That person was a runner, a thin woman dressed in a bright yellow jacket and matching shorts. At some point in the trail, past the patch of mud and wildflowers and after I'd already turned back from the wrong parking area, we crossed paths. I moved to the side so that she could keep her pace, hurrying past me. She nodded at me as she ran by.

The trail narrowed as it continued upward and finally broke through the trees to a high plain, full of brush and gray and orange rocks, sometimes small and flat, sometimes massive boulders that looked like they'd been tossed by large machinery. I walked for an hour or two before I realized the path I was on had tapered, from a wide lane, almost to a trickle, barely enough room for both of my feet to pass. As I looked around, I didn't remember the scattering of rocks or the knee-high bushes or even the tops of the trees blanketing the view; it all looked like something I might see in Wyoming, but not like something I had seen this day or other days on the same trail.

I walked around for a few minutes, circled the path to see if there was something wider in the distance or if I could find my way back to the place I'd come from, but one narrow trail just led to another. After a while I sat down on a large flat rock and tried to remember where I'd seen that runner, what the route was that I'd taken since that last clear spot. The sky was already darkening and I worried—on that rock—what I'd do if I had to spend the night out in the park, with only light clothing and half a bottle of water, my phone and wallet locked in my car at the first parking lot where I'd ventured from.

Reasoning that the park wasn't that large, that if I just wandered, I might find a road, I finally got up and started walking. After a few minutes, I heard the heavy steps of a herd of cows and then bells, likely strapped to their necks. I walked on, followed the sound, and as I turned a bend, there was a patch of wildflowers that looked strangely familiar. Just past them was a single cow standing next to a wide, stony trail.

## XIII

I did not keep the professor's grass alive. It started small: a few patches of brown in the side yard, between the professor's house and the neighbor's. I hadn't walked into that part of the yard for a few days when I noticed a six- or seven-inch section of grass that was a lighter color than the rest, the individual blades of grass spiky and dry. This was the phase before the end, though I didn't realize it at the time, before the grass would lose its ability to stand at all, would lie flaxlike on the hard ground.

I sprayed the spot first with a hose—hoping for a quick resuscitation—and then I set up a long metal sprinkler on that part of the lawn. It watered all around, clicked several times, changed directions and watered again. I got up each morning, turned on the sprinkler and moved it from one position in that side yard to another. I placed the same sprinkler around the yard once in the afternoon and, for good measure, right after dinner or whenever I returned home in the evening.

After the sprinkler came fertilizer, at the suggestion of a guy working at the local hardware store a few blocks away. "Catch it before the rest of the grass wilts," he told me, "before things get out of hand." Wanting anything but the grass to wilt, to get *out of hand*, I hauled the ten-pound bag of fertilizer he'd recommended back to the house and borrowed a friend's spreader to add it evenly to the lawn. I walked in carefully overlapping rows from one side of the yard to the other, the small green spreader rolling along, casting a steady spray of tiny gray pellets across everything.

Still, the spots spread from the side yard to the sunniest corner of the backyard and then even to patches out front, under the large spindly evergreen near the living room window. By the time the professor finally moved back into her house, most of the yard had changed color, no longer a bright green, instead a faded and fading one.

## XIV

"Midway in our life's journey, I went astray from the straight road and woke to find myself alone in a dark wood. How shall I say what wood that was! I never saw so drear, so rank, so arduous a wilderness! Its very memory gives a shape to fear. Death could scarce be more bitter than that place! But since it came to good, I will recount all that I found revealed there by God's grace."

I read these lines to my students one class as we talk about the way the center of the world is not entirely what Axel and Lidenbrock expect. I think about them as I walk after class to the small Bavarian-styled Episcopalian church in the middle of downtown, as I kneel on the kneeler, pray with the eight or nine other people attending that Wednesday service for the peace of the world, as the priest anoints my head with oil for healing, in the name of the Father, of the Son, and of the Holy Ghost.

It's archetypal, the finding one's self alone in a dark wood, the finding one's way back out. "Here's a secret," I will read around this same time in an article. "Everyone, if they live long enough, will lose their way at some point. You will lose your way, you will wake up one morning and find yourself lost." The writer continues, though, his reflection not ending just with losing one's way. "If you're lucky, you'll remember a story you heard as a child, the trick of leaving a trail of breadcrumbs, the idea being that after whatever is going to happen in those woods has happened, you can then trace your steps, find your way back out. But no one said you wouldn't be changed by the hours, the years spent wandering."

## XV

I remember telling a friend when I first visited Laramie that the place had a foreboding sense about it, that it felt, when you drove in—and even when you stayed the night—like an abandoned ghost town that you stumbled into and then realized, "Hey! There are other people here, and those other people are just as surprised to see you."

It was the wide streets that made me feel this way at the time, I think, the occasional tumbleweeds actually blowing through them, and perhaps also that everything had been buried in several feet of snow shortly before I'd gotten there, and so people were mostly inside, or perhaps out in the mountains, skiing and snowshoeing and making the most of it. Some of the neighborhoods around campus—especially the neighborhood where the professor's house was—were actually quite similar to the rows of Craftsman and Victorian houses in the city I'd lived in before coming to Laramie, even the white house with the two-story porch that held my apartment. I didn't see that on my first day in Laramie, though.

I took a photograph my first day driving into town. I'd passed the old cement plant, where arsenic dust would be found sometime later and then remedied by being covered in plastic tarps. I'd driven by the Albany County Fairgrounds and the red-and-white block-lettered sign for Bart's Flea Market. I'd driven up Third Street, past a cigarette stand, two pawn shops; I'd just made it to Grand Avenue, one of the two main thoroughfares between downtown and the rest of the city. To get to the sandstone buildings and pine-tree-lined paths of the university, you turn right at the intersection, but I'd accidentally turned left, toward downtown and West Laramie. As I made that left-hand turn, for some reason, I pulled out my phone and pressed the small camera button on it, taking a picture of what was straight in front of me.

The corners of that photograph include the curved edge of my car's hood. There are white flecks covering the glass and the view and a single black windshield wiper, just off-center on the inside right of the frame.

### XVI

Some years after my summer in Wyoming, I will read an article headlined "Supervolcano? The Big One in Wyoming?" The article details a rift that opened up somewhere near Yellowstone National Park the size of several football fields. It caused scientists to examine again the thirty-by-forty-five-mile caldera that is the impetus for the park's geysers and hot springs.

The rift, the writer notes, is evidence that perhaps the volcano lying some miles under the park's surface may break through. If that volcano blew, the writer of that article, or maybe another, explained, it could be a magnitude eight explosion; its cloud would cover an area five hundred miles wide with ash, perhaps four inches thick.

I look at a map as I read, and I think about the fact that five hundred miles means the volcano would coat all of the park and the surrounding counties and states; it would make its way to Casper and Gillette and even to the city of Laramie, to the tree-lined neighborhoods and the stucco house, and Prexy's Pasture on campus and maybe the Snowy Mountains.

*Would it have mattered*, I wonder as I read, *knowing about the volcano, knowing about the deer tick, knowing that no matter how hard I tried I could not keep the professor's grass green?*

### XVII

To fall in a dream might mean to lose control. Or it might just be waking up, the mind resisting sleep, the brain confusing a signal for rest with one for falling, reminding the body that it must take care, that it must stay alive.

Near the end of *Journey to the Center of the Earth*, Axel, Lidenbrock, and Hans, riding on a small raft, fall over what feels like an endless waterfall. "Greater," Axel narrates, "than the great cataracts of America," with a "surface like a shower of arrows, launched with utmost force... a network of moving threads," going at a speed, he predicts, of ninety miles an hour.

Axel gives himself up for lost. "It will be understood," he says, "that these ideas passed vaguely and darkly through my mind. It was difficult to think at all in this giddy journey which was so like a fall. Judging by the wind that lashed our faces, our pace must have far exceeded that of the most rapid train. To light a torch under these conditions would have been impossible."

What Axel doesn't expect is that the descent is not the end. Instead, Verne sees to it that the protagonists rise, that they do indeed go somewhere. Moments after that waterfall, the small raft that carries Axel, Lidenbrock, and Hans begins to rise and rise, though what those characters rise out of is the shaft of an erupting volcano.

## XVIII

On one of my last nights in Wyoming, my friends and I pulled the car over on the side of the road on the way home from the mountains. We'd stopped in Centennial, a small town at the foot of Medicine Bow National Forest, talked about getting pie at the cedar-lined Bear Tree Tavern, and maybe we did, but then we drove on, past the handful of trailers and cabins, past the land one of our friends would soon buy. We drove down Snowy Range Road toward Laramie, talking and laughing and listening to music on the car radio, watching the green lights from the dials and the dashboard spill onto the floor, until someone in the backseat said, "Stop" or "Look." I can't remember; what I do remember is whichever thing they did say, we stopped and looked.

My friend parked in the uncut grass on the side of the empty road, and we all climbed out of the car. It was silent, and while our eyes adjusted, it was dark—totally dark—no streetlights or lit windows or city haze in the distance, no oncoming headlights or neon bar signs, no glare from the snow and no reflection of the moon off water. It was just black, the sort of black that might have been pure, that might have been annihilating, that meant it took several seconds before we could make out the faces of our friends or the lines of the road or follow the outlines of our own hands.

But then, as we waited there in the middle of the road, our eyes still open, we could see just above the black. Even in that great unknown, there were stars, a canopy.

# ON TIME

### DECEMBER 2015

On December 3, 2015, approximately eighty tons of ice were brought in large refrigerated shipping containers from the Davis Strait near Greenland to Paris, France. There were twelve pieces of ice in total, the largest of which, the *New Yorker* noted, was about the size of three New York taxi cabs "piled on top of each other." Each of the pieces were floaters, breakaways from a larger ice sheet, cast off to sea. Divers and dockworkers from the Royal Arctic Line had collected those floaters, roped them behind ships, and dragged them to shore. Once at the shore, cranes loaded the ice into metal shipping containers, which were ferried by boat to Aalborg, Denmark, and then transferred to trucks, to be driven roughly thirteen hours, likely through Belgium and to France.

When the trucks pulled into Paris, they may have passed Le Marais or Notre Dame or the Musée Curie; they almost certainly crossed the Seine, making their way between the Garden of Luxembourg and the Botanical Gardens through the fifth arrondissement—the Latin Quarter—to the very center of the brick-lined Place du Panthéon, the church-turned-mausoleum that was once dedicated to Saint Genevieve, the patron saint of Paris. There, in view of the columned building, the trucks stopped and deposited the twelve pieces of ice, left them in a twenty-meter circle, each positioned like the hour marker on a clock or on the face of a watch.

It was less than a month after the Paris terrorist attacks had taken place when those trucks pulled into France full of ice, and it was the week of COP 21, the United Nations Conference on Climate Change, the same conference where the Paris Agreement would be negotiated. The blocks of ice were art, the joint project of Icelandic artist Olafur Eliasson and University of Copenhagen natural history professor Minik Rosing. It was Eliasson and Rosing's second installation of ice; the first had gone to Denmark the year before. "As an artist," Eliasson was quoted as saying then, "I am interested in how we give knowledge a body. What does a thought feel like and how can knowledge encourage action?" Later, in "Ice, Art, and Being Human," an essay on the website for the project, Eliasson and Rosing would write together that "one of the challenges of our time is that people feel disconnected from—perhaps even insensitive to—the world's great problems. We do not see ourselves as agents in a global society."

For nine days Greenland and France were inextricably connected as Paris hosted the installation that was those great, dripping pieces of ice. In videos posted after the fact, tourists and locals take photographs with and of the ice. They touch its sheer edges. One man, wearing a green vest, does the robot between blocks of ice. Cars move on the road just behind the blocks; once, a siren is heard going off in the distance. Somebody places a small plastic cup under one of the pieces of ice; in the video footage I see, the cup is about to overflow from catching all the drippings. The installation, *Ice Watch*, was a dual nod, Eliasson noted, to the shape that the pieces of ice made and to the way climate change itself is a race against the clock.

One thousand pieces of ice the same size as the Paris blocks—noted one reporter around that same time—melt every second in Greenland.

**SEPTEMBER 2014**

Just over a year before the Paris Ice Watch, I moved to Michigan, a place where both water and time would soon be on everyone's minds.

Earlier that year, as a cost-cutting measure, the city of Flint changed its water source from Lake Huron and the Detroit River to the Flint River. Almost immediately after the change, city residents began to notice that their water was different. Rather than the formerly clear stream, the water out of the tap was turning brown or orange. A General Motors plant in town observed that its engine parts were beginning to corrode after being washed in city water. In one particularly heated meeting at Flint's city hall, residents brought bottles and even gallons of brown water taken directly from their taps and hoisted them above their heads, calling for an explanation.

Boiling orders came intermittently, but still the city assured residents that the water was fine. The mayor of Flint, Dayne Walling, drank tap water on television to prove to residents that it was sanitary. "Are you ready to drink it?" asked Collette Boyd, the TV-5 news anchor. "Yes, my family and I drink it every day," replied the mayor, before taking a big drink from a white mug. The mayor was young, dressed in a blue power suit with a maroon and tan striped tie.

When a team from Virginia Tech comes to study the water—at the request of a concerned mother—they find lead levels more than a hundred times past the legal limit. Still, it's not until twenty-one months after the city water is switched that a state of emergency is declared: twenty-one months of drinking and showering and bathing babies in that water.

Twenty-one: I will notice later how much the time will come up in discussions of Flint, in articles criticizing the city and the state and even the nation's response. *What if it had only been twenty-one days?* I muse to a friend. *Twenty-one hours? Twenty-one minutes of lead? Would it, should it, still matter then?*

## DECEMBER 2015

Two days before eighty tons of ice were dropped in the shape of a watch at the Place du Panthéon in Paris, technology and finance giant Bloomberg created its own clock. This was a digital clock broadcast on Bloomberg's website. In the background is an image of the earth's surface, clouds floating above the water and then gradually darkening to black. On top of that background, a series of white block numbers quickly tick upward.

The clock reads 409.49617940 when I first I see it; it's a carbon clock, tracking historical carbon increases, since the United States started measuring airborne carbon dioxide in 1958. That year, in 1958, there were 316 parts per million of carbon dioxide in the atmosphere; now, sixty-one years later, there are 409 parts per million. According to Bloomberg, the "danger zone," when sea levels will rise significantly and extreme weather will affect economics and the environment in drastic ways, is 450 parts per million. Bloomberg estimates that the world will reach this danger zone by the year 2040.

Just below the ticking clock, Bloomberg also includes a graph showing likely parts per million of carbon dioxide in the atmosphere from thousands of years ago until today. Until the 1900s the line holds steady, mostly horizontal, usually somewhere below 280 parts per million. Between the 1900s and now, the line indicating global rises in carbon usage becomes nearly vertical.

**FIFTH CENTURY BCE**

In Ancient Greece, there were two words for time: *chronos* and *kairos*. Chronos—pictured as an old man with a gray, flowing beard—referred to chronological time, the basic movement of days and minutes and seconds. Kairos, on the other hand, is defined as "the right time": an opportune time, a crucial time, a sort of seizing of a perfect fleeting moment. In art, the Greek figure for Kairos has often been portrayed as young, beautiful, and standing on his tiptoes, with winged feet ready to run. He's also known for having a single lock of hair on the front of his head only, signifying that you can catch and seize Kairos if you see him coming up, but you have no way of holding onto him if you only see him from behind. In one marble relief of Kairos, the figure holds a pair of scales in one hand and a razor in the other, suggesting, it seems, that Kairos is found somewhere between danger and opportunity.

I call a Greek American friend to ask her about Kairos and, ironically, catch her in the middle of a Greek event in Houston. She confers with the woman sitting next to her and then says she can't tell me much beyond one absolute fact; in Modern Greek, the word *kairos* means weather.

## JUNE 2018

When I visit friends in Oregon, they tell me that they have just started watching a series of films and TV shows coming out of Norway, called "slow films." The films are about various things—train rides, boat rides, campfires, even knitting—and are "slow" because they follow whatever they're filming in real time. Viewers watch a boat drift up a fjord at the real speed it's going and in the moments it's going there; they see each stop the train makes or the person knitting one stitch and then row at a time.

"It's mesmerizing," my friend Sarita tells me. We're on a hike, through an old-growth mountain forest. My friends, used to the climb, are moving steadily upward at a good clip; I'm struggling to keep up behind.

"Slow sounds great," I answer her back, hoping she takes my meaning doubly.

My other friend, Sarita's sister Luisa, chimes in to explain that apparently the shows got so popular in Norway that people who had been watching started to come out and meet the trains and boats, to wave as they went by or to welcome them in to their final destination. "Wouldn't that be fun?" she asks.

I will think later, when I'm looking for those slow films online, about Andy Warhol's iconic 1964 film *Empire*, which showed more than eight hours of black-and-white slow-motion footage of the Empire State Building. One year before that, his film *Sleep* showed five hours of the poet John Giorno sleeping. The films are "conceptual," I had read Warhol saying in an interview; by that I assumed he meant they were to exist in the world more than to be watched. The thing is, though, they are watched, and their slow time or real time does something to the viewer, says something to us about time.

For a few years before my friends tell me about the Norwegian slow films, I'd watched what would likely be classified as a slow film myself around Christmastime. "Yule Log" was a video of a fireplace that burns and crackles. There was something about the noise in the background, constant and steady, that I liked, and also the film seemed to make my tiny apartment seem more festive, like the fireplace was transported through my television screen.

It's privilege, I realize only later, that allows me to watch those slow films, that makes me think I have some control over time.

## JULY 2018

In summer 2018 I come across another web-based clock; this clock is hosted by the University of Oxford and shows a rise in greenhouse gases, moving toward a trillion tons of carbon released into the atmosphere. The clock was designed after a series of papers were published detailing the idea that cumulative carbon emissions are more important than year-by-year tallies.

The website for this clock is simple: a gray-and-white background with big red block numbers ticking away in a frenetic motion. When I first glance at the clock, the number on it reads 622,892,427,000. Only a minute or two later the number has changed to 622,892,437,000. I go back to the site the following afternoon—not even twenty-four hours later—and the number is 622,916,403,967. I read in a write-up of the site that in 2013 the Oxford researchers predicted one trillion tons of carbon dioxide would be reached by Wednesday, November 14, 2040. Not quite five years later, this date has already been moved up to Saturday, April 5, 2036.

I try to watch the clock, even for just ten minutes, to see how it moves, but the speed with which the numbers change is literally dizzying; it begins to hurt my eyes.

## NOVEMBER 2015

Shortly after I move to Michigan, a new friend tells me about a public art installation class, Civic Studio, in which class members collectively design a major art project that comments on the nature of contemporary society. One Civic Studio, for instance, built an "Inconvenience Shop" where a bakery had once existed, raising all sorts of questions about commodification and about changing neighborhoods in our medium-sized Michigan city. Other Civic Studios got at environmental questions and questions of space, land use, and architecture.

The Civic Studio my friend tells me about is a Provisional Flood Club. In one photo I see of the space, there's a boat hoisted into the air with the word "Club" painted in white capital letters on its side. In another photo a square Plexiglas container held up by red crates is partly filled with water. "If you peered down and looked through it," my friend told me, "you could see downtown Grand Rapids rising above floodwaters."

The art installation took the history of the area—a flood zone—and asked how local citizens could pull together if the waters rise again. "Everyone is already a member of the Provisional Flood Club," reads a statement class members put out, "based on the inherent and shared connectedness as inhabitants of the earth and in recognition of water as a fundamental and vital aspect of life." In another statement, summarizing what the Provisional Flood Club aims to be, they write, "As the climate changes, gentrification occurs, and feelings of anxiety and loneliness set in more intensely and frequently, the sense of vulnerability, urgency, and precarity that flood brings will play a larger role in collective imaginaries. Perhaps, as the Provisional Flood Club poses, this vulnerability will not be that which is our downfall, but it will be that which unites us across cultural, ideological, and class divides."

*Is it our civic responsibility*—I will wonder after hearing about that exhibit—*to be a provisional flood club? What if the flood is already here? What if we're told the flood isn't happening, that it's all just ordinary water?*

## NOVEMBER 2016

Exactly one year after my new friend tells me about the Civic Studio, environmental discourse in the United States changes. First, references to climate change are wiped from government websites. Soon after that, the leader of the Environmental Protection Agency says he would not agree that carbon dioxide "is a primary contributor to the global warming we see." Soon after that, the EPA's Clean Power Plan begins to be dismantled. Soon after that, the United States exits the Paris Climate Agreement. Soon after that, implementation of new ozone standards is delayed, first for a year but possibly indefinitely. Soon after that, a federal flood-risk management order is revoked. Soon after that, climate change is taken off the list of concerns to national security. Soon after that, car emissions regulations are loosened, and rules on coal and methane are rolled back. Soon after that—nineteen days into a government shutdown—Florida senator Marco Rubio tries to dissuade the president of the United States from declaring a federal emergency in order to build a wall by saying, "We have to be careful about endorsing broad uses of executive power. Tomorrow the national emergency might be climate change."

## JULY 2006

Nine years before the Paris ice clock, I travel to the glacier—Norway's Jostedalsbreen Glacier. It's my first year in Norway and my first time on the glacier. I'm with my friends Darrell, Annie, CJ, and Jessica; I can't remember who else is with us. I'm wearing all black: long sleeves, long pants, and gloves, low-cut hiking boots, and a full-body red climbing harness. I know this from a photograph someone took and that I still have on my desk. In that photo, Annie is on the right and Jessica on the left. They're both taller than me and both are wearing sunglasses. A rope and a carabiner are tied to Jessica's harness. In the foreground, where we stand, is a short, flat stretch of broken rock. Behind us are several long washes of snow; two icy lakes are visible in the distance, and behind those are three mountains, all snow-covered. Two of the mountains are neatly ridged toward tablelike tops, and one juts out in sharp, jarring edges.

It was a beautiful day that first time on the glacier; you can see in the photograph that my face is reddened, probably from both wind and sun. The landscape was unlike anything I'd ever seen. I don't remember many of the particulars, but I do remember climbing up over a section of blue ice and suddenly having a view of mountains from every angle. I took a photograph of that too: blue skies—bluer than I'd ever seen—set against white, crusted snow, and mountains upon mountains.

On our way home from the glacier, back on the sunny bus some of us had ridden to get there, we all stripped off jackets and gloves and boots. As I was in the habit of doing—in order to get off the fitted sleeves of my hiking overshirt—I unstrapped my watch and put it in the back of the seat pocket in front of me with my gloves and my hat. When the bus stopped at the place where we were getting off, I grabbed the hat and the gloves but somehow missed the watch; it would be some hours before I realized it was gone.

It was my favorite watch—a gift from my parents—one I'd worn through two summers leading canoe, kayaking, and backpacking trips, through another summer as a camp counselor, and another where I'd moved to Colorado. The watch was one of several things I left or misplaced that first summer in Norway. Perhaps I'd simply packed too much in my oversized backpack, or perhaps it was just carelessness. I worried that I would not come back to Norway, that I would return to life in the States and forget about what I had seen there; perhaps I was leaving my own lost possessions like a trail to lead me back.

The thing is I did return to Norway. Less than a year later I boarded the same bus—with different friends this time—to and from the same blue ice. It was on my way home that it happened. I walked down the aisle and sat down near the back. Like the year before, I began unzipping my fleece and jacket and taking off my gloves. It was then that I saw it there—the purple watch, caught in the netted back pocket of the seat in front of me. I picked it up amazed; the band had faded some, but the digital face was still keeping time.

## DECEMBER 2018

Months pass and then years after I find the recovered watch. Still, I think of that story often. Sometimes I wonder if it's supposed to tell me something about time, maybe the way time marches or drives on, with or without us, no matter what happens around it. Or perhaps it is supposed to be a story about the way that things we've lost sometimes return to us, however unexpectedly.

The truth is, even more than the story of the recovered watch, I can't stop thinking of that photograph with my friends on the glacier. My arms were behind my friends' backs in that photograph; you can see both of my hands, but in those first happy days in Norway the watch is just out of sight.

## JUNE, SIXTEENTH CENTURY BCE

The earliest clocks were not made digitally or designed with metal parts. They were actually made of water. Stone bowls or cylinders were designed with a single tiny hole, large enough to either drain or take in water, drip by drip. As the bowl or cylinder filled—or emptied, alternatively—the water moved past a series of etchings, each mark in the stone indicating an hour.

I don't know where the water came from that filled those clocks, but I read somewhere that eventually mercury was used in water's place. With its propensity to freeze and melt, water couldn't keep entirely accurate time.

Dogger Ba

Canary Islands

✗✗                    ✗
The Bermuda      Ma
  Triangle

Bahamas

# AFTER

# ATLANTIS

39.7024° N, 44.2991° E

## MOUNT ARARAT, TURKEY

When I was still a child, the first story I heard about a flood was the tale of Noah's ark. There was a man in that story—Noah—building a boat, and then groups of animals loading themselves onto that boat two by two. While the land around them flooded, Noah, his family, and all those animals rode out the forty-day storm tucked safely inside their floating wooden home. At the end of it all, the waters receded and a rainbow was given to Noah as a promise that there would never again be a flood that large, one that covered the whole earth.

59.8586° N, 17.6389° E

## UPPSALA, SWEDEN

It would be some time after hearing about the ark before I would read of another flooded land, this time the island of Atlantis. In that story—first recorded by Plato in his dialogues *Timaeus* and *Critias*—there is an ancient island kingdom that is a superpower in its own right. "For at this time, the ocean was passable," Plato would write, "since it had an island in it in the front of the strait that you people call the 'Pillars of Hercules.' This island was larger than Libya and Asia combined, and it provided passage to the other islands for people who traveled in those days." The island that Plato described wasn't content to only keep to its boundaries, however, even as vast as they were. Instead, as Plato recounted, "this power gathered all of itself together, and set out to enslave all of the territory inside the strait, including your region, in one fell swoop."

The island's power got larger and larger as it swept across the region, conquering one people after another. Still, like many of the stories I heard as a child, this superpower couldn't get away with its evil advances forever. As the narrative continues, Athens—one of the regions the island attempted to take over and subdue—rose up and decided to fight for its freedom. Even when deserted by their neighbors, the Athenians fought valiantly and, in a surprising feat of victory, overcame the island power, freeing all the people who had been enslaved. Immediately after Athens won, violent earthquakes and floods began to shake the area. Following what Plato calls "the onset of an unbearable day and night,...the entire warrior force sank below the earth all at once, and the island of Atlantis likewise sank below the sea." In a reverse ending to the story of Noah, the flooded waters didn't reveal land; instead, the whole place was lost into the dark depths of the ocean.

Throughout the years since Plato penned the story of Atlantis, many locations have been proposed as possible homes for the ancient submerged city: Malta, Israel, Canaan, the Bimini Road in the Bahamas, the Canary Islands, the Madeira Islands, the Bermuda Triangle, and Indonesia. Ships, explorers, and mapmakers alike have searched for Atlantis or even made claims that they have found it.

Even Sweden has been proposed as a site of the lost Atlantis. Between 1679 and 1702, Olof Rudbeck, the botanist who planted some of Sweden's most famous gardens, argued that his hometown of Uppsala was the location of the lost Atlantis. In his 2,500-page *Atlantica*, Rudbeck set out to prove his theory, using mythology and etymology as evidence for his over one hundred reasons why Uppsala matched Plato's narrative.

Looking through the online holdings of a rare map store in California, I find an inked-in map commissioned by Rudbeck. Latitude and longitude lines are sketched on the bottom and right side of the map and veer in thin lines across it. Mountains, rivers, and towns are added to the map, along with dozens of small islands following Sweden's shoreline. There, in approximately the spot where Uppsala now stands, is the label "Atlandah."

41.8683° N, 88.0996° W

## WHEATON, ILLINOIS

When I was in graduate school, I lived in a city that occasionally flooded. My whole life before this, I'd never lived in a low place, never had to consider the flow of water. The old farmhouse which I shared with two other roommates in that city, though, was prone to inundation. Bordered in front by a parking lot and in back by railroad tracks, it was out of place just a few buildings from the main thoroughfare to downtown. On the second floor we were fine, but in one storm after another, our neighbors in the downstairs apartment pulled sofas and rugs and bookcases onto our porch to dry. The narrow driveway flooded, and so did the campus parking lot down the street.

Temporary *No Parking* signs appeared during storms, and my neighbors and I usually remembered to move our cars. One weekend when we were all away, though, one of my roommates accidentally left her car in that low-lying campus parking lot. When she returned, there was her lone white car, drowning in two or three feet of water. Though the neighbors' apartment did not flood during that storm, several buildings on campus did. In one dorm the weight of the water caused windows on the first floor to break, forcing students to evacuate into lounges or to stay with friends, some for the rest of the semester. That night a friend of mine who lived in the dorm walked outside and saw lightning strike ground in front of her. Some years later she would tell me that the image never left her, that when she saw that bolt on the campus lawn she thought at first it was a tree on fire.

After the storm was over, black-and-white newspaper photos showed students swimming in the streets. None of us knew that the water was not only coming down from the sky but also up from the sewers.

54.7214° N, 2.7678° E

## THE DOGGER BANK, NORTH SEA

In 1999 French author Jean Deruelle published *Atlantide des Mégalithes*, a book that drew on older Atlantis theories as well as historical and geographical research to propose that a strip of submerged land called Doggerland was the one-time site of Atlantis.

Deep in the North Sea between the UK and Europe, Doggerland was named in the 1990s by British archaeologist Byronny Coles after "doggers" or Dutch fishing boats that used to travel there. Though many meters under the surface now, that land is thought to have once been above water, before massive floods took place there in the middle of the seventh century BCE. Deruelle and others proposed that the floods were triggered by a series of earthquakes that broke open methane deposits on the ocean floor and caused more than 180 miles of Norwegian coastline to collapse into the sea. From that force came a tsunami. The Dogger Bank—and whatever inhabited it—was lost, once and seemingly for all time, beneath the North Sea.

Deruelle believed there was clear evidence that the Dogger Bank fit the Atlantis narrative, most notably because it is one of the world's few submerged places that is home to a "Great Plain," one of Plato's descriptions of Atlantis. In the years before Deruelle wrote his book, trawlers had dragged up some man-made artifacts from the area, further evidence, in Deruelle's mind, that an ancient civilization had once claimed this spot.

In the almost two decades since Deruelle's book, the Dogger Bank and its surrounding area, Doggerland, have become a site of more study. In 2015 a team of researchers from the University of Bradford and several other British universities received a €2.5 million grant from the European Research Council to study and map Doggerland using vast remote-sensing data acquired from oil and gas companies.

59.8586° N, 17.6389° E

## UPPSALA, SWEDEN

Sometime after finding the Swedish Atlantis map in the California map store, I visited Uppsala. On my way, I made a detour in nearby Stockholm to see the Vasa Museum, home of a 1,200-ton wooden Swedish warship that sank in 1628, less than twenty minutes after taking off for its first voyage and only two years before Rudbeck, that originator of the Swedish Atlantis theory, was born.

I took my time walking through the museum, circling all around the bronzed ship, past its tall masts and intricate carvings: a lion with a coat of arms in its paws, two baby angels, the Greek god Hercules, and a series of warriors. While touring exhibits about the people who would have worked on the ship and packed or handled the artifacts on display, I found myself most interested in the narrative of the sinking—a top-heavy boat catching the wind, then taking on water only 1,300 meters from shore, while the crowd of onlookers who'd come to see the maiden voyage watched, unable to help. Walking through that museum, I wondered whether the sinking of the *Vasa* was on the edges of Rudbeck's consciousness when he began to study Atlantis, whether the idea of something being not lost but drowned, or maybe buried, led to his revisionist theories on Plato's island.

Later that day I boarded a train for Uppsala, an old and impressive university city with smoky cafés, streets packed with bicycles, and a narrow river running through downtown. I visited the university, a local castle, and the red brick Uppsala Cathedral, the site where Rudbeck is buried. I wandered through the gardens that he planted but still could not find anything in all those places that echoed the myth of Atlantis.

My last stop in Sweden was a boat trip through the archipelago, "channels" as Rudbeck called them, borrowing the term from Plato. It was a bright day with a slight wind as the low boat motored away from the shore and headed out of the harbor and then past one after another tiny island, just off the coastline. Mostly the other boats we passed were small, like ours. Sometime midway through the day, though, a reconstructed sailing ship—maybe from the time of the *Vasa* or the doggers—sailed by, the body of the ship dark wood with a deep-red mast and a Dutch flag billowing in the sea air.

I try now to imagine, as I think about that ship, Rudbeck himself standing on the deck of a similar boat, looking out over what he is certain is the ancient Atlantis: not just an island, but a country.

14.5994° S, 28.6731° W

## ATLANTIC OCEAN

When I began to study the story of Atlantis, I assumed the place was named after the Titan god Atlas who, after getting on the wrong side of a war, was condemned to forever hold up the sky, the same god collections of maps—atlases—are named after. It turns out that there are two Greek gods named Atlas. The Atlas who ruled Atlantis wasn't a Titan; he was the son of Poseidon, the god of the sea. Perhaps his end shouldn't have been surprising, given his parentage. Perhaps it's also not surprising that the name "Atlantic Ocean" is tied to this god and his submerged home.

Of course, Plato's Atlantis isn't just a story of a lost city; it is a cautionary tale too, a moral about what happens when a place gets too powerful and tries to rise up against its neighbors.

There's one other thing I learned upon revisiting Plato. Some scholars consider *Critias* to be the first environmental text; in it Plato also laments the destruction of forests near Athens. Many great floods, he explains, have eroded the soil and changed the landscape until the city has become like "the bones of the wasted body."

36.3932° N, 25.4615° E

## SANTORINI, GREECE

Jean Deruelle and Olof Rudbeck are not the only ones to argue Atlantis might have been real. Many historians and scholars believe that there could have been a historical basis for Atlantis, even if the stories that have come down to us are fictionalized. As in the case of the Dogger Bank, these theories often hinge on cataclysmic events.

Perhaps even more likely than the earthquakes of Doggerland is the eruption of a volcano in Santorini, Greece, sometime between 1600–1500 B.C.E. Not only did the lava and ash consume whole cities, it's thought that a tsunami followed, submerging nearby coastal areas. In the 1960s, an ancient Minoan settlement was discovered beneath the ash of the volcano on Santorini. Pottery and furniture and frescoes began to emerge; soon, three-story buildings appeared. It seemed newly possible that this site—buried and not that far geographically from Plato's Athens—may have been the basis for Plato's story. Seventy miles away, on the island of Crete, scattered artifacts discovered along the coast seem to back up this theory, or at least the idea that a series of very large waves had hit land there too.

A few historians line up the Santorini eruption with the Biblical story of Exodus. They see that great migration of the Israelites from Egypt as coinciding not only with the Old Testament story of the Israelites fleeing slavery under Pharaoh, but also with some unforeseen natural disaster. Others have said maybe the myth of Atlantis is evidence, however distant, validating the story of Noah and the flood, Atlantis being what was left in Plato's day of the memory of that long-ago event.

41.8683° N, 88.0996° W

## WHEATON, ILLINOIS

My classmates and I called that year in graduate school when everything flooded "the year of the plagues." It had started with mumps; that fall we had gotten an email that a case of mumps had been discovered at our small college. Unfortunately, the mumps-carrying student didn't realize what they had and so had gone to a couple of classes and the library and eaten in the dining hall. Only a few weeks later there were already thirty-some cases on campus. Infected students were quarantined in their rooms, and student workers were tapped to leave trays of food outside their doors.

After the mumps came the cicadas—masses of them—a hum in the background of every conversation. They had emerged as part of a seventeen-year cycle, coming out of the ground in droves. One local newspaper noted that each cicada's call was equal to the decibel noise of a kitchen blender. People worried about the effect on outdoor weddings that spring; a local music venue moved the date for its yearly classical concert so the cellos and violins wouldn't have to compete with the droning.

Before the cicadas left, the roof of a campus building caught on fire while some students were still inside. A blowtorch had been left on by construction workers above the main auditorium. The workers had gone home before anyone realized what had happened. Soon the roof started smoking. Campus safety officers and firemen rushed to clear the building; the auditorium was closed for the rest of the year.

Finally there was the flood.

Our plagues weren't the same as those of the Old Testament—boils, locusts, hail, a river turned into blood—but they were close enough to make us wonder.

29.9511° N, 90.0715° W

## NEW ORLEANS, LOUISIANA

In 2005, a year and a half before the campus plagues, the levees broke and New Orleans flooded. Hurricane Katrina had hit that August, beginning in the Bahamas and making its way over Florida, across the Gulf of Mexico, and up to Louisiana. New Orleans—with Lake Pontchartrain to its north, Lake Borgne and the Mississippi sound to its east, the Gulf of Mexico to its south, and the Mississippi River winding through it—was supposed to be protected by a series of levees and floodwalls. Within twenty-four hours of Katrina's landfall, one then two then five then ten then fifteen then twenty of those levees and floodwalls gave way to the water. When all was said and done, at least fifty levees and floodwalls had failed, leaving 80 percent of the city sitting under water.

The news showed video footage of people being rescued from their roofs and of roads full of drowned cars. In aerial shots you could see only the tops of buildings and houses in one neighborhood after another; all the streets and driveways and yards, and the first floors of many buildings, were completely underwater. The city's disaster plan for major storms like Katrina was unearthed sometime later. It warned that there could be "thousands of fatalities," waterborne and airborne toxins, even "floating coffins," and that people should be evacuated at least three days before a high-category storm hit land. At the time of Hurricane Katrina, evacuation orders were issued just twenty-four hours ahead.

29.7604° N, 95.3698° W

## HOUSTON, TEXAS

In 2017, ten years after the plagues, Hurricane Harvey descended and the city of Houston flooded. My friends posted videos online, from the vantage point of their suburban front porches, of the water rising. In one of the videos, a friend stood inside her house with the front door open. The water had filled the entire street, had covered the yard, and was lapping up, not far from the doorway where she stood. The video panned out, showing the water across the street or what was at one time a street but was now a flat, wet expanse. The water rose past the yards and seeped into the houses. Just as my friend stopped filming, a man in a long red kayak paddled by.

One old friend from college posted about boarding up her windows for the first time and driving north. A former running partner bought a quick plane ticket out of state and figured she would sort out lodging and food when she arrived. Several other friends and acquaintances heeded the mayor's warnings not to leave town and crowd the highways. Instead, they stayed put, hoping their houses wouldn't fill with water.

34.2104° N, 77.8868° W
## WILMINGTON, NORTH CAROLINA

There is a science behind real and recently submerged cities. As temperatures rise across the globe, glaciers, icebergs, and icecaps melt, draining water into the ocean. All that ice has to go somewhere, whether into the air or into the water. It's not only what's added, though; warming water expands, and so even a small surge in temperatures, when spread out over an ocean's worth of water, can mean significant rises in water levels, threatening many low-lying regions. Since the 1880s, sea levels have risen eight inches globally. If temperatures continue to rise at their current rate, by 2100 the earth will be 3.2°C warmer than it was in the preindustrial era, raising ocean levels anywhere between one and six feet and putting Shanghai, Osaka, Alexandria, Rio de Janeiro, Amsterdam, London, Miami, Oakland, and other cities at risk. In the next twenty years some of these cities are predicted to become victims of "chronic, disruptive flooding" or flooding that occurs biweekly on average and impacts at least 10 percent of the city's land.

Warming temperatures also mean more water vapor, feeding

hurricanes and increasing their wind strength and intensity. In a March 2017 article in the *Bulletin of the American Meteorological Society*, MIT atmospheric sciences professor Kerry Emanuel details the results of six thousand storm simulations. Factoring for projected changes in temperature, Kerry predicts that by the end of this century, there will be a 118 percent uptick in storms that increase by sixty knots, or about seventy miles an hour, in the twenty-four hours before they hit land. In the past there was about one of these storms every hundred years. By the end of the twenty-first century, they will happen every five to ten years.

Only two years after Hurricane Harvey came Hurricanes Irma, Maria, Florence, and Michael. After Houston came San Juan, Puerto Rico; Jacksonville, Florida; Wilmington, North Carolina; and Hilo, Hawaii.

43.4501° N, 87.2220° W

## LAKE MICHIGAN

Not long after Hurricane Harvey, I have a dream that my students and I are traveling to attend a literary festival at a park in the middle of a deep woods. Dark evergreens line the road, making everything feel shadowy and hushed. We're riding in a van that one of the students is driving. The dream is weird in the way dreams often are; it's the first day of April and also Easter, and part of the dream involves taking strange turns and not having enough cash for gas or entrance fees to the park and having to stop repeatedly until we find some.

It's early in the morning, still dim, when we finally get to the park. We pile out of the van, all of us standing together in a small clearing. As we look around, no one else is there: no booths or speakers or books for sale, no one waiting in lines and no greeters telling us where to go. Instead, in every direction, even the one we've come from, the ground emits smoke. Before any of us has time to decide exactly what to do, the park has become an island of small fires: our own Atlantis, burning instead of sinking.

In real life, sometime before that dream, I did take my students on a trip, but it wasn't to a dark forest. It was to one of the Great Lakes. Our university owned a small tugboat, the *D. J. Angus*, that was used as a science research vessel. I'd checked the weather ahead of our trip and cautioned my students to dress warmly, to wear layers and a rain jacket, and to bring anything that would help them take in the journey: pens, notebooks, cameras, binoculars if they had them. In the end, the trip was actually blisteringly hot, and some of my students, delayed after stopping at McDonald's, almost didn't make it aboard.

As we set off, two scientists told us what to watch for and talked about the history of the lake and the fish, plants, and animals residing in it. They detailed the release of Asian carp into the lake and the problematic zebra mussels that had made their way there in the water tanks of ships. My students took furious notes and one photo after another. Before we were beyond the bay, where we launched, and into the wilder, darker waters, one student happened to see a shard of wood sticking out of the water.

"What is that?" she asked.

"When ships weren't useful anymore," one of the scientists replied, "they were burned or sunk or left just offshore. Since the waters have risen now, all we can see of them are the masts. What you see there, that's the body of a ship."

Long after the boat trip I will think about that leftover body, mostly submerged in water. I will wonder what it looked like before it was sunk. I will wonder what other bodies are left under the water and what bodies are to come.

25.0847° N, 77.3235° W

## ATLANTIS RESORT, BAHAMAS

One spring, after a particularly long winter, my family and I visited the Atlantis resort in the Bahamas. We were just there on a day trip, a quick stopover, before continuing on elsewhere. I hoped, as soon as we started making travel plans to Nassau, that we might catch a boat out to the ocean, perhaps near the Bimini Road, an ancient underwater rock path that some Atlantis theorists believe is Plato's sunken land. It rained hard that full day, though, so instead we took a van from the main city center to the resort.

Long before the van pulled up to the site, we could make out the resort's Royal Hotel, a castlelike, salmon-colored building topped by turrets and with a large enclosed bridge, several stories high, connecting its two main buildings. My family and I walked through the hotel, past its lobby and aquariums with swimming schools of brightly colored fish, past the onsite casino and high-end shops: Gucci, Rolex, Versace. We walked down palm-tree-lined paths, by pools and waterslides built to look like ancient stepped-stone ruins. We saw a marine habitat called the Dig, where groupers, seahorses, eels, and jellyfish swim among statues and pottery shards—a mockup of Plato's ancient island city. When a break came in the rain, we headed toward the beach. We walked through scores of empty sunbathing chairs and by an outdoor café, blaring music to only a couple of people sitting at the bar. When we finally reached the edge of the water, the kids ran and played in the white sand while we talked about maybe swimming if the rain continued to hold off.

In the end, the rain started pouring again. We grabbed the towels and kids and hurried toward the nearest shelter, an overhang beneath a small, artificial lagoon.

Just for a moment, in the crowd of people that somehow materialized and was also rushing to take cover, I lost my family. There were people everywhere, crowding into the dimly lit stone structure. It turns out it was the Predator Lagoon, the home of Atlantis's sharks and barracudas. As I searched the crowd for familiar faces, thunder boomed overhead and I heard a single, piercing scream. The floor was wet, maybe from all the people's feet or maybe from incoming rain. But what I couldn't get out of my mind, in that moment, searching for my family in the Predator Lagoon, was a vision of the waters continuing to rise, that small trickle becoming a stream, the stream becoming waves and then rushing in so fast that none of us would know what had happened.

40.7462° N, 14.4989° E

## POMPEII, ITALY

As a child, I was fascinated by the stories of lost worlds: *Treasure Island*, Narnia, even the history of Pompeii, buried and then found after so many years. In graduate school I got caught up in maps, which took me back to the lost world, the mapped and unmapped: *Erewhon, King Solomon's Mines, Gulliver's Travels, Coral Island, Journey to the Center of the Earth*, and finally Atlantis. I wanted, more than anything, to travel to these places, to find them in real life or at least to live within their boundaries as long as I could on the page.

It was the idea that you could walk through the back of a wardrobe or sail just past what you know and find yourself somewhere else. The lost worlds were brilliant and illusory and proof, it seemed, of the value of keeping your eyes wide open. The thing is, as a child I never considered that the world lost at sea might be our own or the one we'd left behind, let go until it was swallowed whole.

82.8628° S, 135.0000° E

## ANTARCTICA

In 1995 two librarians from British Columbia, Rose and Rand Flem-Ath, put forth another site for the historical Atlantis. Building on the theories of Charles Hapgood, one of Einstein's contemporaries, the Flem-Aths argued that twelve thousand years ago the earth's crust shifted, forcing the continents into new positions and causing earthquakes, volcanic eruptions, and floods. It was amid all this upheaval that the ancient island of Atlantis was cast off into the sea.

The Flem-Aths' book, *When the Sky Fell*, proposed that Atlantis's real home was Antarctica's Ross Ice Shelf; rather than lost underwater, Atlantis was buried beneath all that ice.

# NOTES

## BAFFIN ISLAND

The major source for the Norse myth of the Ginnungagap is Snorri Sturluson's thirteenth-century *Prose Edda*. I also consulted English translations of the *Prose Edda*, including the Penguin Classics, Oxford World Classics, and Everyman's Library editions.

PAGE 3. *chaos of perfect silence*  Daniel McCoy, founder of Norsemythology.org.

PAGE 3. *yawning emptiness*  Several writers use this term in reference to the Ginnungagap. I first came across it in Amy T. Peterson and David J. Dunworth, *Mythology in Our Midst: A Guide to Cultural References* (Greenwood Press, 2004).

PAGE 4. *one entitled* Norse Stories  Hamilton Wright Mabie, *Norse Stories Retold from the Eddas*, edited by Katharine Lee Bates (Rand McNally, 1902).

PAGE 5. *The first map I find*  I consulted several sources on the relationship between cartography and the Ginnungagap: Kristen Seaver, *Maps, Myths, and Men: The Story of the Vinland Map* (Stanford University Press, 2004); Fridtjof Nansen, *In Northern Mists: Arctic Exploration in Early Times* (Frederick A. Stokes, 1911); William Babcock, *Legendary Islands of the Atlantic: A Study in Medieval Geography* (American Geographical Society, 1922); Halldór Hermannsson, *Islandica*

(Cornell University Press, 1920); and James Robert Enterline, *Erikson, Eskimos, and Columbus: Medieval European Knowledge of America* (Johns Hopkins University Press, 2002).

PAGE 7. *a single flaming sword* From Snorri Sturluson, *Prose Edda*, translated by Jesse L. Byock (Penguin Classics, 2004).

PAGE 8. *Swedish visual artist Sigrid Sandström* Photographs of Sandström's exhibit are at inmangallery.com/artists/sandstrom_sigrid/2004_Ginnungagap/13.html and fryemuseum.org/exhibition/72.

PAGE 8. *When a man plants a flag* Jen Graves, "Wondering about Wandering: Sandström Asks What We Expect to Find Out There," *The Stranger*, June 22, 2006.

PAGE 9. *The Atlas Universel* Martin West, "The Mystery of the Vaugondy Maps," *Western Pennsylvania History* (Summer 2001).

PAGE 9. *a technology of knowledge* Anne McClintock, *Imperial Leather: Race, Gender, and Sexuality in the Colonial Contest* (Routledge, 1995), 27–28.

PAGE 11. *Researchers led by Gifford Miller* "Cause and Onset of the Little Ice Age," University of Colorado Institute for Arctic Research, instaar.colorado.edu/research/projects/cause-and-onset-of-little-ice-age.

PAGE 12. *Buenos Aires, Argentina* Sources for short stories about maps and mapmakers in this section include Jorge Luis Borges, "Del rigor en la ciencia" ("On Exactitude in Science"), *Los Anales de Buenos Aires* (1946); Umberto Eco, "On the Impossibility of Drawing a Map of the Empire on a Scale of 1 to 1," in *How to Travel with a Salmon and Other Essays* (Harcourt, 1994); and Neil Gaiman, "The Mapmaker," in *Fragile Things: Short Fictions and Wonders* (Harper, 2010). I also consulted Jorge Luis Borges's translation of the *Prose Edda*, *La alucinación de Gylfi* (Alianza Editorial, 1984).

## THEORY OF WORLD ICE

PAGE 20. *Austrian engineer Hanns Hörbiger* Robert Matthias Erdbeer introduced me to Hörbiger in his spring 2013 lecture at the University of Missouri, "Counter-Science: The World Ice Movement's Cosmic Visions and Its Rise to Public Fame (1894–1945)." To Erdbeer, Hörbiger's theory of world ice "was not only a theory of the universe;

it was also a universal theory" accounting for biology, geology, the arts, and the humanities. Walter Gratzer's brief discussion of Hörbiger in *The Undergrowth of Science: Delusion, Self-Deception and Human Frailty* (Oxford University Press, 2000) provides a succinct explanation of Hörbiger's theory of the cosmos centered on ice.

PAGE 22. *All ice is made out of hydrogen and oxygen atoms* The ice facts in this section are from Kenneth Chang, "Explaining Ice: The Answers Are Slippery," *New York Times,* February 21, 2006.

PAGE 25. *one time an elephant was led across the river* Tom de Castella, "Frost Fair: When an Elephant Walked on the Frozen River Thames," *BBC News Magazine,* January 28, 2014.

PAGE 25. *exorcists were brought in* "History of Chamonix Glaciers," Chamonix.net.

PAGE 25. *Professors were molested in the streets* Louis Pauwels and Jacques Bergier, *The Morning of the Magicians* (Dorset Press, 1988).

PAGE 26. *Modern Science seems to foster a desire for a final synthesis* Max Benzen, 1934, cited by Erdbeer, "Counter-Science."

PAGE 26. *Frederic Tudor of the Tudor Ice Company, was the first* See Christopher Klein, "The Man Who Shipped New England Ice around the World," *History,* August 29, 2012; and Rebecca Rupp, "Frederic Tudor: The King of Ice," *National Geographic,* July 19, 2014.

PAGE 29. *Glaciers in Norway have begun to creep down* Alister Doyle, "In Norway, Glaciers Are Growing Bigger," *Los Angeles Times,* October 28, 1990.

PAGE 31. *the astronomy of the invisible* Christina Wessely, "Cosmic Ice Theory: Science, Fiction and the Public, 1894–1945," Max Planck Institute for the History of Science, www.mpiwg-berlin.mpg.de/research/projects/DeptIII-ChristinaWessely-Welteislehre.

## THE PHILOSOPHER'S CABIN

Primary sources for details on Wittgenstein include Ludwig Wittgenstein, *Tractatus Logico-Philosophicus,* translated by D. F. Pears and B. F. McGuinness (Routledge, 2014); Bertrand Russell, *The Collected Papers of Bertrand Russell,* edited by John G. Slater, vol. 8 (Routledge, 1986); Ludwig Wittgenstein et al., *Letters to Russell, Keynes, and Moore* (Cornell University Press, 1974); Ludwig Wittgenstein, *Wittgenstein in Cambridge: Letters and Documents, 1911–1951*, edited by Brian McGuinness (Wiley-Blackwell, 2012); Paul Engelmann, *Letters from Ludwig Wittgenstein, with a Memoir* (Basil Blackwell, 1967); David Pinsent, *A Portrait of Wittgenstein as a Young Man: From the Diary of David Hume Pinsent, 1912–1914*, edited by G. H. von Wright (Basil Blackwell, 1990); and Ludwig Wittgenstein, *Philosophische Untersuchungen: Philosophical Investigations*, edited by P. M. S. Hacker and Joachim Schulte (Wiley-Blackwell, 2009).

Secondary sources include *The Oxford Handbook of Wittgenstein*, edited by Oskari Kuusela and Marie McGinn (Oxford University Press, 2011); "Wittgenstein: The Philosopher and His Works," Publications of the Austrian Ludwig Wittgenstein Society, New Series 2, edited by Alois Pichler and Simo Säätelaä (2006); Bernhard Leitner, *The Wittgenstein House* (Princeton Architectural Press, 2000); *Wittgenstein and His Times: Essays by Anthony Kenny, Brian McGuinness, J. C. Nyiri, Rush Rhees, and G. H. von Wright*, edited by Brian McGuinness (Basil Blackwell, 1982); and *The Cambridge Companion to Wittgenstein*, edited by Hans Sluga and David G. Stern (Cambridge University Press, 1996).

Skjolden resident Edvin Bolstad supplied additional information about Wittgenstein and his history in the village.

## DRIVING WYOMING

Most of the information on Japan's volcanoes comes from the Global Volcanism Program, volcano.si.edu. Also, unless directly noted, all italicized Craig Arnold quotations come from his volcano travelogues, available at volcanopilgrim.wordpress.com.

PAGE 61. *perpetual presence of the sublime*   From Ralph Waldo Emerson's 1836 essay "Nature."

PAGE 66. *Pompeii…Buried Alive!*   Edith Kunhardt, *Pompeii…Buried Alive!* (Random House, 1987).

PAGE 67. *at least forty people have died*   When I returned to look for the original newspaper source regarding deaths on 287, I was unable to find it in my files. Another similar map was created in 2017, though, by Amelia Arvesen, "US 287 Backdrop for 4 of 19 Boulder County Fatal Crashes in 2017," *Longmont Times-Call*, September 23, 2017.

NOTES                                                                     255

PAGE 68. *a crash killed eight University of Wyoming student athletes* "8 UW Cross-Country Athletes Killed in Crash," *Billings Gazette*, September 16, 2001. For updated information on Highway 287 crashes being twice as likely to be fatal, see Jeremy Pelzer, "Wyoming Officials: No Easy Solutions to U.S. Highway 287's High Fatality Rate," *Casper Star Tribune*, September 12, 2010.

PAGE 69. *The moon is brighter* Bashō quotations come from *Bashō's Journey: The Literary Prose of Matsuo Bashō*, translated by David Landis Barnhill (State University of New York Press, 2005).

PAGE 74. *an empty space or interval; interruption in continuity* Italicized rephrasing of the word "gap" is a direct quotation from Dictionary.com.

PAGE 75. *"Hymn to Persephone"* Craig Arnold, *Made Flesh* (Copper Canyon Press, 2008).

PAGE 75. *paralyzing horror* W. G. Sebald, *The Rings of Saturn* (New Directions, 1998).

PAGE 75. *The Lady and the Monk* Pico Iyer, *The Lady and the Monk: Four Seasons in Kyoto* (Vintage, 1991).

PAGE 76. *the mechanism of terror* Joan Didion, *Salvador* (Simon & Schuster, 1983).

## LOST: AN INVENTORY

PAGE 79. *Lost Jewels* Angelique Chrisafis, "Mont Blanc Climber Finds £205,000 Worth of Indian Jewels on Glacier," *Guardian*, September 26, 2013.

PAGE 80. *Lost in Ice* "On Ice 4,000 Years, Bronze Age Man Is Found," *New York Times*, September 26, 1991; "50,000-Year-Old Plant May Warn of the Death of Tropical Ice Caps," *Ohio State News*, December 11, 2004, news.osu.edu/news/2004/12/11/quelplant; Laura Spienney, "Melting Glaciers in Northern Italy Reveal Corpses of WWI Soldiers," *Telegraph*, January 13, 2014; Rolleiv Solholm, "Old Shoe—Even Older," *Norway Post*, December 4, 2016; "Ancient Tool Found in Melting Ice Near Yellowstone," *Wall Street Journal*, June 29, 2010; "1959 Message in a Bottle a Clue to Glacier Melt," *CBC News*, December 6, 2013.

PAGE 82. *Lost: Biology and Psychology* Paul Dudchenko, *Why People*

*Get Lost: The Psychology and Neuroscience of Spatial Cognition* (Oxford University Press, 2010); K. A. Hill, *Lost Person Behavior* (National SAR Secretariat, 1998).

PAGE 86. *Lost on Mont Blanc*   Lane Wallace, "Why Is Mont Blanc One of the World's Deadliest Mountains?" *Atlantic*, July 25, 2012; "Cleaning Up the Piste: Europe's Highest Toilets Installed on Mont Blanc," *Spiegel Online*, July 24, 2007.

PAGE 86. *Lost in Mont Blanc*   Percy Bysshe Shelley, "Mont Blanc: Lines Written in the Vale of Chamouni," in *History of a Six Weeks' Tour through a Part of France, Switzerland, Germany, and Holland: With Letters Descriptive of a Sail Round the Lake of Geneva, and of the Glaciers of Chamouni* (T. Hookham and C & J Ollier, 1817).

PAGE 86. *Lost Promise*   "Statement by President Trump on the Paris Climate Accord," White House, June 1, 2017, whitehouse.gov/briefings-statements/statement-president-trump-paris-climate-accord.

PAGE 87. *Lost Place*   Paul Voosen, "Delaware-Sized Iceberg Splits from Antarctica," *Science*, July 12, 2017.

PAGE 87. *Lost Science*   Lisa Friedman, "E.P.A. Cancels Talk on Climate Change by Agency Scientists," *New York Times*, October 22, 2017.

PAGE 87. *Lost Map*   "Antarctic Dispatches," *New York Times*, May 18, 2017.

PAGE 88. *Lost Father*   Nick Pisa, "Three Now in Court Fight over Who Found Iceman Case," *Edmonton Journal*, October 23, 2005; "Iceman's Discoverer Dead in the Alps," *BBC News*, October 23, 2004; Kathy Marks, "Science vs. Superstition: Curse of the Iceman," *Independent*, November 5, 2005; Stephen Goodwin, "Obituary: Helmut Simon; Finder of a Bronze Age Man in the Alpine Snow," *Independent*, October 25, 2004.

PAGE 90. *Lost Valley*   Srinivas Laxmani, "Homi Bhabha: Operative Spoke of CIA Hand in 1996 Crash," *Times of India*, July 30, 2017.

PAGE 91. *Lost in Ice*   Patrick Bodenham, "The Mystery of Mont Blanc's Hidden Treasure," *BBC News*, March 14, 2014; Simon Johnson, "Letters Lost in Plane Crash Recovered after 60 Years," *Telegraph*, July 6, 2010; Ruth Doherty, "Preserved Hand of 1966 Plane Crash Victim

Found on Mountain," AOL News UK, aol.co.uk/2017/07/31/preserved-hand-plane-crash-victim-51-years-ago-alps-mountain; Henry Samuel, "French Treasure Hunter Finds 50 Pieces of Jewelry on Mont Blanc from Air India Crash 48 Years Ago," *Telegraph*, August 17, 2014.

PAGE 94. */lôst, läst/* "Am I losing it?" italki Answers—Learn English, www.italki.com/question/267253?hl=en-us.

## GLACIOLOGY

Two key sources not specifically mentioned are Charles S. Elton, *Voles, Mice, and Lemmings: Problems in Population Dynamics* (Clarendon Press, 1942); and Nils Chr. Stenseth and Rolf Anker, *The Biology of Lemmings* (Linnean Society of London / Academic Press, 1993).

PAGE 98. *Some scientists say that the Norwegian lemming deaths* Walter Sullivan, "Scientists Find a Clue to Mysterious Mass Death of Lemmings," *New York Times*, May 7, 1969.

PAGE 98. *it's a stress mechanism* Garrett C. Clough, "Lemmings and Population Problems," *American Scientist* 53, no. 2 (1965).

PAGE 98. *In the 1960s, scientist W. B. Quay* Kai Curry Lindahl, "New Theory on a Fabled Exodus," in *The Natural History Reader in Animal Behavior*, edited by Howard Topoff (Columbia University Press, 1987).

PAGE 98. *It is mass hysteria* Ivan T. Sanderson, "The Mystery of Migration," *Saturday Evening Post*, July 15, 1944.

PAGE 101. *In the 1950s, Walt Disney's* White Wilderness *White Wilderness*, directed by James Algar (Walt Disney Productions, 1958).

PAGE 102. *it wasn't the year of the lemmings* "Cruel Camera," hosted by Bob McKeown (CBC Television, 1982), youtube.com/watch?v=DG4jnhrSukQ.

PAGE 102. *Norwegian lemmings are the only* W. A. Fuller, "Hinterland Who's Who: Lemmings," Canadian Wildlife Federation, hww.ca/assets/pdfs/factsheets/lemmings-en.pdf.

PAGE 102. *They fall from the sky* Kai Curry Lindahl, "New Theory on a Fabled Exodus," in *The Natural History Reader in Animal Behavior*, edited by Howard Topoff (Columbia University Press, 1987).

PAGE 102. *Collared lemmings* Agnes Kruchio, "Lemmings: Creatures Are Kingpins, Not Cliff-Jumpers," *Globe and Mail*, February 4, 1980.

PAGE 104. *Initially scientists believed the Norwegian lemmings* B. L. Wick, "Norwegian Lemming," *The Friend*, February 29, 1896.

PAGE 104. *in the 1990s, biologists began to notice* Eliza Strickland, "Norway's Lemming Populations Plunge Off the Statistical Cliff," *Discover*, November 5, 2008.

PAGE 104. *lemming fossils from the Pleistocene* Vadim B. Fedorov and Nils Chr. Stenseth, "Glacial Survival of the Norwegian Lemming (*Lemmus lemmus*) in Scandinavia: Inference from Mitochondrial DNA Variation," *Proceedings of the Royal Society: Biological Sciences* (2001).

PAGE 104. *Biologists began to say it isn't the cold* Rols A. Ims et al., "Determinants of Lemming Outbreaks," *Proceedings of the National Academy of Science* 108, no. 5 (2011).

PAGE 104. *The stress from the overcrowding and hunger* Garrett C. Clough, "Lemmings and Population Problems," *American Scientist* 53, no. 2 (1965).

PAGE 105. *The lemming population is falling* Alister Doyle, "Lemmings in Norway Hit by Global Warming," Reuters, November 5, 2008.

PAGE 106. *following his death in December 1966* Leonard Mosley, *Disney's World: A Biography* (Scarborough House, 1990).

PAGE 106. *Disney's family and friends* David Blatty, "Disney on Ice: Walt's Frosty Afterlife?" *Biography*, December 15, 2014, biography.com/news/walt-disney-frozen-after-death-myth.

PAGE 106. *I realize the tragic significance* Harry S. Truman, "Radio Report to the American People on the Potsdam Conference," August 9, 1945; retrieved from Harry S. Truman Presidential Library and Museum, trumanlibrary.org/publicpapers/?pid=104.

PAGE 107. *scientists began to date glacial accumulation* Ohio State University, "New Tibetan Ice Cores Missing A-bomb Blast Markers; Suggest Himalayan Ice Fields Haven't Grown in Last 50 Years," *Science Daily*, sciencedaily.com/releases/2007/12/071211232938.htm.

PAGE 107. *Our films have provided thrilling entertainment* Jim

NOTES                                                                                         259

Korkis, "Behind the True-Life Cameras," Walt Disney Family Museum, wdfmuseum.org/blog/behind-true-life-cameras.

PAGE 107. *In his 1976 children's book*   Alan Arkin, *The Lemming Condition* (Harper & Row, 1976).

PAGE 107. *In a 2008 bid for the Senate*   Lemmings ad, Andrew for Oklahoma, youtube.com/watch?v=KShl3xFwUVs.

PAGE 111. *set out to analyze*   Natalie Kehrwald et al., "Mass Loss on Himalayan Glacier Endangers Water Resources," *Geophysical Research Letters* 35, no. 22 (2008).

## ABOUT THE COLLECTION

PAGE 113. *the Breheimsenteret was conceived*   The Breheimsenteret website is at jostedal.com/en/breheimsenteret.

PAGE 114. *In contrast to the souvenir*   Susan Stewart, *On Longing: Narratives of the Miniature, the Gigantic, the Souvenir, the Collection* (Duke University Press, 1992).

PAGE 115. *systematic collection and study of evidence*   Jeffrey Abt, "The Origins of the Public Museum," in *A Companion to Museum Studies*, edited by Sharon Macdonald (Wiley-Blackwell, 2011).

PAGE 116. *The thrill that students*   D. V. Proctor, "Museums: Teachers, Students, Children," in *Museums, Imagination, and Education* (UNESCO, 1973).

PAGE 118. *Without close attention*   Archaeological Institute of America, www.archaeological.org.

PAGE 118. *was designed to facilitate an encyclopedic enterprise*   Ashmolean Museum of Art and Archaeology, ashmolean.org.

PAGE 119. *lemmings do not fall from the sky*   The Victoria and Albert Museum "Born on This Day" series, which in May 2015 featured Ole Worm.

PAGE 124. *The container and the contained*   Ada Louise Huxtable, "Art and Architecture as One," *Wall Street Journal*, October 16, 1997.

PAGE 124. *full of the promise of aesthetic and poetic power*   Ada Louise Huxtable, "Northern Enclosure: Hot Museums in a Cold Climate," *Wall Street Journal*, May 14, 1998.

PAGE 125. *The museum's origins*   The Ashmolean offers a history of John Tradescant and Elias Ashmole at ashmolean.org/history-ashmolean.

PAGE 129. *Downright near infinite*   Lawrence Weschler quotation, from the jacket of *Infinite City: A San Francisco Atlas*, by Rebecca Solnit (University of California Press, 2010).

PAGE 129. *To build a museum*   Sverre Fehn quoted in Per Olaf Fjeld, *Sverre Fehn: The Pattern of Thought* (Monacelli Press, 2009). Also, *Architect: The Work of the Pritzker Prize Laureates in Their Own Words* (Black Dog & Leventhal, 2010) is an excellent source on Fehn and his glacier museum.

PAGE 129. *In 1973 Swedish museologist*   Ulla Keding Olofsson, "Temporary and Travelling Exhibitions," in *Museums, Imagination, and Education* (UNESCO, 1973).

PAGE 133. *Freakish winter wildfires in Norway*   Andy McElroy, "Winter Wild-Fires: Norway's New Risk," February 5, 2014, United Nations Office for Disaster Risk Reduction, www.unisdr.org/archive/36397.

PAGE 133. *You feel like you are seeing*   Craig Arnold, volcanopilgrim.wordpress.com.

## PAULING'S CORE

Sources for Linus Pauling include the Pauling Blog (paulingblog.wordpress.com); a September 1935 photograph of Linus Pauling and Norman Elliot, Oregon State University Libraries Special Collections and Archives Research Center, Corvallis; *Linus Pauling in His Own Words: Selected Writings, Speeches, and Interviews*, edited by Barbara Marinacci (Simon & Schuster, 1995); and a biography of Pauling on the Nobel Prize website, nobelprize.org/nobel_prizes/peace/laureates/1962/pauling-bio.html.

Sources for Barclay Kamb include "Barclay Kamb," International Glaciological Society, www.igsoc.org/news/kamb; "Antarctic Landmarks Named after Caltech Experts on Glacier Ice Flow," California Institute of Technology Press Release (2003); "W. Barclay Kamb: Caltech Glaciologist Instrumental in Antarctic Ice Stream Studies," *Antarctic*

*Sun*, May 6, 2011, antarcticsun.usap.gov/features/contenthandler.cfm?id=2422; "W. Barclay Kamb, 79," caltech.edu/news/w-barclay-kamb-79-1915; and Hermann Engelhardt, "Barclay Kamb: 1931–2011," National Academy of Sciences (2018), nasonline.org/publications/biographical-memoirs/memoir-pdfs/kamb-barclay.pdf.

PAGE 134. *It has been generally recognized*   Linus Pauling, "The Structure and Entropy of Ice and Other Crystals with Some Randomness of Atomic Arrangement," *Journal of American Chemical Society* 57, no. 12 (1935), 2680–84.

PAGE 135. *a flash of insight*   Thomas Hager, *Force of Nature: The Life of Linus Pauling* (Simon & Schuster, 1995).

PAGE 135. *Seeing everything*   Algis Valiunas, "The Man Who Thought of Everything," *New Atlantis: A Journal of Technology and Society* 45 (Spring 2015).

PAGE 137. *The study of ice began*   Chester C. Langway Jr., "History of Early Polar Ice Cores," *U.S. Army Corps of Engineers Report* (2008).

PAGE 138. *Ice studies gradually moved away*   J. Jouzel, "A Brief History of Ice Core Science over the Last 50 Years," *Climate of the Past* 9, no. 6 (2013).

PAGE 140. *largest depository of ice outside of nature*   "National Science Foundation Ice Core Facility," icecores.org/about/index.shtml.

PAGE 144. *on January 30, 1960, Linus Pauling*   "Scientist Saved," *Seattle Times*, February 1, 1960; "Pauling Saved from Coast Cliff," *New York Times*, February 1, 1960; "Dr. Pauling Rescued, On Sea Cliff 24 Hrs," *New York Herald Tribune*, February 1, 1960.

## THE ICEBERG PROPOSAL

The Emirates Iceberg Project video is available at youtube.com/watch?v=CdC3HPNjc94. Several journalists note that one problem with the video is that there are polar bears on top of the iceberg that has been dragged to the United Arab Emirates, and polar bears don't live in Antarctica. I agree that this is a problem, though maybe not the largest.

Sources mentioned or consulted on the project include "A United Arab Emirates Company Wants to Tow Icebergs from Antarctica to Combat Drought," *Science Alert News*, May 13, 2017; Samuel Osborne,

"Icebergs to Be Towed from Antarctica to United Arab Emirates for Drinking Water," *Independent*, May 3, 2017; Lara Pearce, "UAE Company Wants to Tow an Iceberg from Antarctica for Drinking Water," *Huffington Post*, May 18, 2017; "Dubai Wants to Drag Icebergs from Antarctica for Fresh Water," *New York Post*, May 17, 2017; Callum Paton, "Why Moving Icebergs from Antarctica to Dubai to Harvest Water Is So Difficult," *Newsweek*, May 17, 2017; and Ian Sample, "Could Towing Icebergs to Hot Places Solve the World's Water Shortages?" *Guardian*, May 5, 2017.

Also consulted were "Iceberg Facts," *Canadian Geographic*, March 1, 2016, canadiangeographic.ca/article/iceberg-facts; Pål Prestrud, "Why Are Snow and Ice Important to Us?" Center for International Climate and Environmental Research, wedocs.unep.org/bitstream/handle/20.500.11822/14473/GEO_C2_LowRes.pdf; and a photograph of a seminar at the Scripps Institute of Oceanography, held in the George H. Scripps Memorial Marine Biological Building, 1950, UC San Diego Library Special Collections and Archives.

PAGE 148. *When I started to optimize it* Daniel Behrman with John Isaacs, *John Isaacs and His Oceans* (American Geophysical Union, 1992).

PAGE 150. *the original minutes for one part of this conference* "History of SOLAS," International Maritime Organization, imo.org/en/KnowledgeCentre/ReferencesAndArchives/HistoryofSOLAS/Pages/default.aspx; original 1913–1914 SOLAS conference minutes, International Maritime Organization.

PAGE 151. *still 2.3 iceberg crashes a year* Lauren Everitt, "Titanic Threat: Why Do Ships Still Hit Icebergs?" *BBC News Magazine*, March 20, 2012.

PAGE 152. *first recorded use of the phrase "tip of the iceberg"* Julia Cresswell, *Oxford Dictionary of Word Origins* (Oxford University Press, 2010), and "Origin and Meaning of Iceberg," *Online Etymology Dictionary*, etymonline.com/word/iceberg.

PAGE 152. *Hemingway began to talk about "iceberg theory"* Ernest Hemingway, *Death in the Afternoon* (Scribner, 1996).

PAGE 154. *understand the effects of flipping icebergs* J. C. Burton et al., "Laboratory Investigations of Iceberg Capsize Dynamics, Energy Dissipation, and Tsunamigenesis," *Journal of Geophysical Research* (2012).

NOTES  263

PAGE 154. *same amount of energy as an atomic bomb*  Stephen Ornes, "Flipping Icebergs: Capsizing Icebergs May Release as Much Energy as Bomb," *Science News for Students*, April 3, 2012.

PAGE 154. *one of these upside-down icebergs*  Melissa Wiley, "An Iceberg Flipped Over and Its Underside Is Breathtaking," *Smithsonian*, January 22, 2015.

PAGE 155. *by means of pipes and air-pumps*  Alexis Madrigal, "The Many Failures and Few Successes of Zany Iceberg Towing Schemes," *Atlantic*, August 10, 2011.

PAGE 159. *It's no big deal*  Michael Ryan, "Iceberg Wrangler," *Smithsonian*, February 2003.

## THE SPEED OF FALLING

The information on Galileo is primarily from the Museo Galileo in Florence, Italy. Other sources include Galileo's *On Motion* and Aristotle's *Physics*. Edvin and Brit Bolstad told me the history of Fuglesteg.

## CAIRNS

PAGE 181. *would even leave messages in cairns*  Michael Gaige, "A Natural and Social History of Cairn Building and Maintenance," *Appalachian Mountain Club Outdoors*, March/April 2013.

PAGE 181. *Cairns embody and invoke*  Paul Basu, "Cairns in the Landscape: Migrant Stones and Migrant Stories in Scotland and Its Diaspora," in *Landscapes beyond Land: Routes, Aesthetics, and Narratives* (Oxford University Press, 2012).

PAGE 185. *People in England didn't believe Rae*  "Cannibalism and Cover Up: Why History Spurned Orkney's John Rae," *Scotsman*, July 20, 2006.

PAGE 186. *The note in the cairn was updated*  David Williams, *Cairns: Messengers in Stone* (Mountaineers Books, 2012).

PAGE 187. *It would fill my office*  Aislinn Sarnacki, "Acadia's Rock

Pile Experiment Noted in New Book," *Bangor Daily News*, December 12, 2012.

PAGE 187. *His rock statue* Strangler Cairn    Priscilla Frank, "Andy Goldworthy's Ephemeral Rock Sculpture 'Strangler Cairn' Cost Australian Tax Payers Almost $700,000," *Huffington Post*, August 29, 2012.

PAGE 188. *witness trees*    Kent C. Ryden, *Mapping the Invisible Landscape: Folklore, Writing, and a Sense of Place* (University of Iowa Press, 1993).

PAGE 190. *what's now known as the Ross expedition to Antarctica*    "Early Explorers: James Clark Ross," *Antarctic Guide*, antarcticguide.com/about-antarctica/antarctic-history/early-explorers/james-clark-ross.

PAGE 190. *We were startled*    Robert McCormick, *Voyages of Discovery in the Arctic and Antarctic Seas and around the World: Being Personal Narratives of Attempts to Reach the North and South Poles*, vol. 2 (Cambridge University Press, 1884).

PAGE 193. *it wasn't until 1938 that two hundred*    "National Tourist Routes in Norway: Sognefjellet," Norwegian Public Roads Administration Brochure.

PAGE 194. *read me a story about cairns*    C. S. Lewis, *The Silver Chair* (Macmillan, 1953).

PAGE 195. *it wasn't until 2014 that the HMS Erebus*    Simon Worrall, "How the Discovery of Two Lost Ships Solved an Arctic Mystery," *National Geographic*, April 16, 2017.

## TO THE CENTER

The two translations I quote from for Jules Verne's *Journey to the Center of the Earth* are the 1876 George Routledge & Sons edition and the 1871 Griffith and Farran edition. The block quotation in section 9 is from chapter 32 of *Journey to the Center of the Earth* (Routledge & Sons, 1876).

PAGE 215. *Midway in our life's journey*    Dante, *Divine Comedy*.

PAGE 215. *Here's a secret*    Nick Flynn, "The Ticking Is the Bomb," *Esquire*, January 24, 2008.

## ON TIME

The sources I consulted on the Flint, Michigan, water crisis are Merrit Kennedy, "Lead-Laced Water in Flint: A Step-by-Step Look at the Makings of a Crisis," NPR.org; Steve Carmody, "Flint Meeting Eases Few Concerns about Safety of the City's Water," *Michigan Radio*, January 21, 2015; and "Flint Mayor Drinks Tap Water," youtube.com/watch?v=ZQ8VPhY2EoI&feature=youtu.be. Source material in the November 2016 section is from "A Running List of How President Trump Is Changing Environmental Policy," news.nationalgeographic.com/2017/03/how-trump-is-changing-science-environment. Also see John Schwartz and Tik Root, "Could a Future President Declare a Climate Emergency?" *New York Times*, January 16, 2019.

PAGE 220. *piled on top of each other* Cynthia Zarin, "The Artist Who Is Bringing Icebergs to Paris," *New Yorker*, December 5, 2015.

PAGE 221. *joint project of Icelandic artist Olafur Eliasson* Nina Azzarello, "Olafur Eliasson Moves 100 Tonnes of Ice to Copenhagen to Visualize Climate Change," *Designboom*, designboom.com/art/olafur-eliasson-ice-watch-project-climate-change-greenland-copenhagen-10-24-2014.

PAGE 221. *one of the challenges of our time* Olafur Eliasson and Minik Rosing, "Ice, Art, and Being Human," icewatchparis.com.

PAGE 223. *Bloomberg created its own clock* Bloomberg carbon clock, bloomberg.com/graphics/carbon-clock.

PAGE 226. *This clock is hosted by the University of Oxford* University of Oxford, trillionthtonne.org.

PAGE 226. *Oxford researchers predicted* Sophie Yeo, "How Long until the World Emits Its Trillionth Tonne of $CO_2$," *Climate Home*, October 28, 2013.

PAGE 227. *Provisional Flood Club* Civic Studio, facebook.com/events/278553392308396 and civicstudio.org/content/provisional-flood-club.

## ATLANTIS

PAGE 236. *the ocean was passable* Plato, *Timaeus and Critias*, translated by Desmond Lee (Penguin Classics, 2008).

PAGE 237. *Rudbeck set out to prove his theory* Olof Rudbeck, *Atlantica* (Uppsala Press, 1679).

PAGE 238. *In 1999 French author* Jean Deruelle, *Atlantide des Mégalithes* (France-Empire Editions, 1999).

PAGE 238. *Dutch fishing boats that used to travel there* Byronny Coles, "Doggerland: The Cultural Dynamics of a Shifting Coastline," Geological Society of London Special Publications (2000).

PAGE 239. *In 2015 a team of researchers* Sarah Knapton, "British Atlantis: Archaeologists Begin Exploring Lost World of Doggerland," *Telegraph*, September 1, 2015.

PAGE 242. *a series of very large waves* Harvey Lilley, "The Wave That Destroyed Atlantis," *BBC News*, April 20, 2007.

PAGE 242. *A few historians line up the Santorini eruption* John Noble Wilford, "New Find Is Linked to Events of Exodus," *New York Times*, December 24, 1985.

PAGE 245. *If temperatures continue to rise* Josh Holder, Niko Kommenda, and Jonathan Watts, "The Three-Degree World," *Guardian*, November 3, 2017.

PAGE 246. *Factoring for projected changes* Kerry Emmanuel, "Will Global Warming Make Hurricane Forecasting More Difficult?" *Bulletin of the American Meteorological Society* (2017).

PAGE 250. *the Flem-Aths argued that* Rand and Rose Flem-Ath, *When the Sky Fell: In Search of Atlantis* (St. Martin's, 1997).

## ACKNOWLEDGMENTS

Thanks isn't nearly enough for Kate Northrop's and S. A. Stepanek's constant kindness, brilliant insights, and friendship (but I'll say it anyway). Elizabeth Chang once dictated a reference for me from a hospital bed, and her thoughts on museums, killer plants, and all sorts of other things fundamentally changed the way I approach my work. Endless gratitude goes to Maureen Stanton, Scott Cairns, Aliki Barnstone, and Julija Šukys, whose careful readings and steady encouragement helped make this project a book. Thanks also to Marguerite Avery and everyone at Trinity University Press for their excitement about and commitment to this manuscript; I'm so glad it found a home there.

I'm lucky to have the best of friends, many of whom have appeared in these pages or in the stories that they tell: Kristin and Will Bankston, Deborah Bates Walker, Claire Donaghey, Heather Brown, Ingvill and Olav Humlebrekke, Kaja Thorjussen, Laura and Andrew Gillott, Callie Flemming, Sam and Maria Aylmer, Steve and Debs McClure, Jane Mitchell, Luisa Gallagher, Sarita Gallagher Edwards, Gez and Alison Perry, Annie Shepherd, Sonya Eikum, Alison Lisiak, Kim Gordon, Becca Hall, Joann

Olson, Phil Cannon, Amanda Kretsinger, Steffi Conradt, Luba Tveiterås, Liz & Daniel, and the Sara(h)s (B, W, and Z). Joanna Eleftheriou consistently talked with me about these essays (and everything else) on trail runs and phone calls and over email; your book is next!

Thanks also to all my friends in the Navs but especially to Lowri Peters, Tom Feather, Luke O'Dowd, Inna Green, Adam and Kate Collett, Annie and Tim Forester, Lydia Clarke, Stephen Hartwell, Cat Thompson, Martha Scott, Emma Connan, Charlotte Caspers, Rachel Jamieson, Tom Candy, Hannah Stewart, Matthew Mellon, Stephen Hamilton (whose name I'm singing as I type this), Korryn Shoge, Rachel Bryan, Lauren Tellman, and the Briners. The Dooyemas first brought me to Norway, and the McClures invited me back (year after year!). I'm grateful to Edvin Bolstad for his work with the Navigators and for his wisdom. Also to Daniel Peterson for joining me in Norway, to Tom and Kris Cleveland for being regular and enthusiastic readers, and to my Crossroads pals, Barb, Joni, and Estee, for their friendship.

I'm indebted to many peers and professors at the University of Wyoming but especially to Paula Wright, Julie Church, Dixie Thoman, Courtney Carlson, Emilene Ostilind, Christina Ingoglia, Evie Hemphill, and Mary P. Sheridan. Beth Loffreda taught me much about what sort of writer I hope to become; Wyoming's MFA was a gift. I'm also grateful to my faculty and to my weekly lunch friends at the University of Missouri (especially Jenni, Alison, and Brie); I'm happy too to have more recently made it onto the conference circuit with Travis Scholl and Corinna Cook (may there be many more panels in our future!).

Thanks also goes to Charles Hanson, Allyson Goldin Loomis, Brett Foster, Scottie May, and Rachel Strom Bass, who were the best sort of teachers, and to Amy Robohm, who was such a help when I most needed it. (I'm back to running!) Rebecca Solnit first mentioned Trinity University Press to me on a visit to Grand Valley State University. Nick Flynn and Lia Purpura read very early drafts of "Glaciology" and "Driving Wyoming" and provided invaluable guidance. My students and colleagues spur me on to more and better writing. Special thanks to Chris Haven, Laurence José, and Amorak Huey for all the great advice, and also to Caitlin Horrocks and Todd Kaneko for welcoming me to Michigan and for the continued welcome ever since.

Finally, thanks go to my parents, to David and Ellen, to Steve and Stacy, and to my four favorite kids: Allie, Maggie, Will, and Adaline Peterson—thanks for asking to hear these stories (again!).

Research for this book was generously funded by a Creative Arts Fellowship from the American-Scandinavian Foundation, a Catalyst Grant from the Center for Scholarly and Creative Excellence at Grand Valley State University, and a Judith A. and Richard B. Schwartz Travel Award from the University of Missouri. I'm also grateful to the editors at *River Teeth*, *Passages North*, *Newfound*, *The Pinch*, *Ocean State Review*, *Flyway*, *Mid-American Review*, *Terrain.org*, and *Post Road* for taking a chance on these essays, and for making them better.

BETH PETERSON is an assistant professor of writing at Grand Valley State University in Grand Rapids, Michigan. Her essays and poetry have appeared in *Fourth Genre*, *River Teeth*, *Post Road*, the *Mid-American Review*, the *Pinch*, *Newfound*, *Passages North*, *Flyway*, *Sky Island Journal*, *Terrain.org*, and other journals. She has an MFA in creative writing from the University of Wyoming and a PhD in literature and creative writing from the University of Missouri.